Environmental Soil Biology

Tertiary Level Biology

A series covering selected areas of biology at advanced undergraduate level. While designed specifically for course options at this level within Universities and Polytechnics, the series will be of great value to specialists and research workers in other fields who require knowledge of the essentials of a subject.

Recent titles in the series:

Title	Author(s)
Social Behaviour in Mammals	Poole
Seabird Ecology	Furness and Monaghan
The Biochemistry of Energy Utilization in Plants	Dennis
The Behavioural Ecology of Ants	Sudd and Franks
Anaerobic Bacteria	Holland, Knapp and Shoesmith
Evolutionary Principles	Calow
Seabird Ecology	Furness and Monaghan
An Introduction to Marine Science (2nd edn.)	Meadows and Campbell
Seed Dormancy and Germination	Bradbeer
Plant Molecular Biology (2nd edn.)	Grierson and Covey
Polar Ecology	Stonehouse
The Estuarine Ecosystem (2nd edn.)	McLusky
Soil Biology	Wood
Photosynthesis	Gregory
The Cytoskeleton and Cell Motility	Preston, King and Hyams
Waterfowl Ecology	Owen and Black
Tropical Rain Forest Ecology (2nd edn.)	Mabberley
Fish Ecology	Wootton
Solute Transport in Plants	Flowers and Yeo
Human Evolution	Bilsborough
Principles and Techniques of Contemporary Taxonomy	Quicke
Biology of Fishes (2nd edn.)	Bone, Marshall and Blaxter

Tertiary Level Biology

Environmental Soil Biology

Second Edition

MARTIN WOOD
Reader in Soil Ecology
Department of Soil Science
The University of Reading

BLACKIE ACADEMIC & PROFESSIONAL
An Imprint of Chapman & Hall
London · Glasgow · Weinheim · New York · Tokyo · Melbourne · Madras

Published by
Blackie Academic & Professional, an imprint of Chapman & Hall
Wester Cleddens Road, Bishopbriggs, Glasgow G64 2NZ

Chapman & Hall, 2–6 Boundary Row, London SE1 8HN, UK

Blackie Academic & Professional, Wester Cleddens Road, Bishopbriggs, Glasgow G64 2NZ, UK

Chapman & Hall GmbH, Pappelallee 3, 69469 Weinheim, Germany

Chapman & Hall USA, 115 Fifth Avenue, Fourth Floor, New York, NY 10003, USA

Chapman & Hall Japan, ITP-Japan, Kyowa Building, 3F, 2-2-1 Hirakawacho, Chiyoda-ku, Tokyo 102, Japan

DA Book (Aust.) Pty Ltd, 648 Whitehorse Road, Mitcham 3132, Victoria, Australia

Chapman & Hall India, R. Seshadri, 32 Second Main Road, CIT East, Madras 600 035, India

First edition 1989
Second edition 1995

Typeset in 10/12pt Times by Cambrian Typesetters, Frimley, Surrey

Printed in Great Britain by The University Press, Cambridge

ISBN 0 7514 0342 3 (HB) 0 7514 0343 1 (PB)

A catalogue record for this book is available from the British Library

Library of Congress Catalog Card Number: 95–77024

∞ Printed on permanent acid-free text paper, manufactured in accordance with ANSI/NISO Z39.48-1992 (Permanence of Paper)

Preface

Soil biology is a theme which runs through many of the major areas of modern science; the subject encompasses global issues concerning the environment, conservation and food production, and the tools used for its study range from molecular biology to the common spade. The aim of this book, which has been largely re-written since the first edition of *Soil Biology* (1989), is to provide an account of the subject for undergraduate and postgraduate students of environmental science and related subjects.

The first part of the book provides an introduction to soils, its inhabitants, and their activities. The second part covers the influence of man on the natural cycles of soil. Topics such as acid rain and nitrogen fertilisers are considered alongside pesticides and genetically modified organisms. A new final chapter has been added which considers how, as we move towards the next millennium, we can apply the concept of sustainability to issues such as global climate change and farming systems.

Much of the work for this book was carried out whilst I was on sabbatical in New Zealand; I am grateful to the staff of the Department of Soil Science at Lincoln University for their stimulating company and hospitality, and to the British Council and the Stapledon Memorial Trust for providing financial assistance.

Finally, thanks not only to Diane, Rebecca and Jennifer, but also to William and to Donald.

M.W.

Contents

1 Soil as a habitat for organisms

1.1 Introduction

Everyone is familiar with soil. We play on it, grow food in it, dump things on it, and occasionally become agitated when we think we might be losing it. However, any ideas of soil being simply dirt or mud are quickly dispelled when we look more closely at this material. The light microscope and the electron microscope reveal a complex arrangement of solids (sand, silt and clay particles and organic matter) and spaces (filled with air and water) as illustrated in Figure 1.1. Within the spaces live plant roots and soil organisms, an incredibly diverse world ranging from microbes to moles. All this makes soil one of the most complex and fascinating of ecosystems.

Scientists who are attracted to studying the biology of soil are therefore faced with many challenges which require an understanding not only of biology but also an appreciation of the principles of physics and chemistry. Inevitably they will find themselves at some time or other despairing over the complexity and variability of the material with which they have chosen to work. Spatial and temporal variability cause problems for all environmental scientists; however, these problems are exacerbated for soil biologists who are faced with the inherent variability of biological systems, together with the fact that the processes of interest often take place in localised favourable areas of the soil which may be anything from a few micrometres to a few millimetres in size. However, it is this very heterogeneity of soil that provides the range of niches which support its great biodiversity.

This chapter sets the scene for the later topics by considering the major physical and chemical features of soil as a habitat for organisms.

1.2 The soil environment

When studying soils it is important to have a clear understanding of the relative magnitude of the different components. It is easy to become absorbed in the fine detail, particularly in the laboratory, and to lose track of the place of soil science in the world around us.

Soils forms only a relatively thin layer over about one-third of the surface of our planet; the roots of crop plants, for example, do not grow much beyond 1 m depth of soil. The growth of such plants depends upon factors such as the way soil solids are arranged to provide channels approximately 0.2 mm in diameter. For their supply of nutrients such as nitrate, plants depend upon the activity of microorganisms such as bacteria which are approximately 1 μm in length. The survival of bacteria in an acid soil may depend upon their ability to exclude or detoxify Al atoms 0.05 nm (5×10^{-5} μm) in diameter. These topics are considered later, but it is important at this stage to be aware of these different scales (Table 1.1).

1.2.1 Inorganic material

Inorganic material comprises the largest proportion of soil (Figure 1.1 and Table 1.2). Sand particles (20–2000 μm diameter) and silt particles (2–20 μm diameter) are mostly quartz, and are relatively inert. Clay particles are composed of clay minerals and often other minerals such as quartz, $CaCO_3$ and Fe and Al oxides. Clay minerals occur in soils as primary minerals or as weathered or re-synthesised secondary minerals.

Table 1.1 The size of some important physical and biological components of soil

Component	Diameter (μm)
Sand particle	2000–20
Plant root	200
Pore (transmission)	>50
Protozoan (ciliate)	20
Silt particle	20–2
Root hair	10
Bacterium	1
Pore (storage)	50–0.5
Clay particle	<2
Pore (residual)	<0.5
Al atom	5×10^{-5}

Table 1.2 The composition, by weight, of a sandy loam (Sonning series) at The University of Reading Farm, Sonning, Berkshire, England

Component	Content (wt %)
Sand	80
Silt	8
Clay	10
Organic matter	2
[pH 6.4]	

They comprise mainly aluminium silicates arranged in plates or lamellae 0.7–1.0 nm thick. The lamellae are stacked together to form clay crystals and these crystals may become oriented along the surface of pores or sand grains, forming clay skins in certain soils.

A thin section of soil, when viewed under a petrological microscope, may show regions of oriented clay lamellae (Figure 1.1), which are often larger than clay particles, termed clay domains. The proportion of sand, silt and clay determines the texture of a soil, which determines the sizes of the solid structures. However, it is the spaces between these solid structures which are of greatest importance to soil organisms; the spaces contain air, water and solutes and serve as the soil's respiratory and circulatory system.

Clay minerals play an additional important role. The size and shape of the clay minerals confer on them a large surface area, most pronounced with the smectite clays, montmorillonite being the most common example (Table 1.3). Furthermore, the substitution of one atom within a clay lamella by another atom of similar size but different charge (isomorphous substitution), for example Al for Si, gives rise to a net negative charge in the clay particle. This charge is independent of the pH of the surrounding soil solution. However, the edges of clays and the surfaces of hydrated Fe and Al oxides have a charge which depends upon pH. As the acidity increases (pH decreases) the surface charge becomes more positive. Carboxylic and phenolic groups in soil organic matter also contribute pH-dependent charge (becoming increasingly negative at pH >3) of 150–300 cmol kg^{-1} organic matter.

The predominantly negative charge on the surface of clays and organic matter attracts positively charged cations. The cations which are adsorbed to these surfaces are freely exchangeable with cations in solution and this property, termed cation exchange (Table 1.3), is important in determining the concentration and movement of nutrients in soil. As a result of these and other properties it has been suggested by Cairns-Smith (1985) that clay minerals played a role in the origin of life, serving as prototype nucleic acids.

Clay minerals can affect microbial activity. These effects appear to be mainly indirect rather than involving direct interactions between soil solids and microorganisms. For example, the stimulatory effects of montmorillonite on fungi and actinomycetes, and the ability of clays to protect microorganisms against the inhibitory effects of osmotic potential are mainly due to the ability of clay minerals to modify the local physical and chemical conditions in the soil. However, microorganisms are not readily leached from soil and are often difficult to dislodge, which indicates that they do adhere to soil components.

Surfaces in soil are not homogeneous; although most of the surface will be clay minerals, there may also be coatings of hydrous metal oxides and

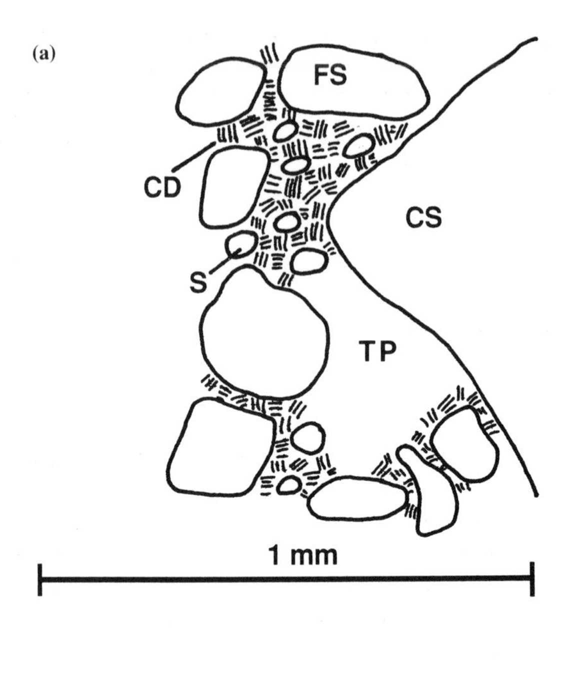

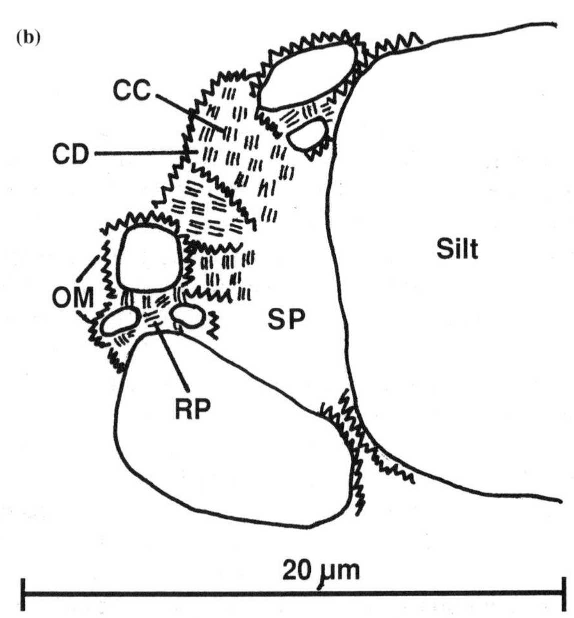

Figure 1.1 A schematic diagram of soil particles as seen at two levels of magnification: (a) light microscope; (b) electron microscope (CC, clay crystal; CD, clay domain; CS, coarse sand; FS, fine sand; OM, organic matter; RP, residual pore; S, silt; SP, storage pore; TP, transmission pore). Courtesy of Dr D.L. Rowell, The University of Reading.

Table 1.3 Surface area and cation exchange capacity (CEC) of soil particles. After White (1987)

Particle	Specific surface ($m^2\ g^{-1}$)	CEC ($cmol\ kg^{-1}$)
Sand	0.01–0.1	0
Silt	1.0	0
Clays:		
Kaolinite	5–100	3–20
Montmorillonite	700–800	100
Vermiculite	300–500	100–150

organic matter. It may be that clay minerals concentrate organic substrates at their surfaces allowing microorganisms to grow in what would otherwise be unfavourable conditions. However, if clay surfaces are enriched with nutrients it is not clear whether these nutrients are available for microbial growth. Clay surfaces may also bind antimicrobial compounds which might otherwise be detrimental to microorganisms, and inactivate certain pesticides, e.g. paraquat.

1.2.2 Organic matter

Soil organic matter is derived from all the organisms which live in or on soil. The variety of organisms (from bacteria to trees) and the vast number of molecules making up these organisms (from simple amino acids to complex aromatic polyphenol polymers such as lignin) together with their decay products, give rise to a highly complex material. The plant and animal residues in soil are not equally susceptible to decomposition by microorganisms; part of these residues remains in soil for a period of time in a modified form, and under certain conditions (e.g. peat bogs) may accumulate. This material is found beneath the surface litter layers of soil as a layer of amorphous brown material in which the identity of the original material is lost. This resistant material is termed humus.

The nature of humus has been the subject of intense scientific investigation for the past 200 years, particularly by chemists, but still today its precise chemical composition remains a mystery. For example, only approximately 50% of the organic N in soil has been identified as compounds such as amino acids, amino sugars and nucleic acids. It is known, however, that humus possesses certain properties which distinguish it from the plant and animal materials from which it has been formed (Table 1.4).

Several pathways have been proposed for the formation of humus (sometimes referred to as humic substances) during the decomposition of

Table 1.4 Some of the major characteristics of humus

Dark brown/black in colour
Plant and animal remains not visible
Virtually insoluble in water
Dissolves in dilute alkali
Certain components dissolve in acid
Contains 55–60% carbon (higher than plant material)
Remarkably constant C:N ratio of 10
High capacity for base exchange
Capacity to adsorb pesticides and other organic chemicals
Capacity to adsorb heavy metals
Modifies the physical and chemical properties of soil e.g. colour, texture, moisture holding capacity, aeration
Determines the composition and activity of the microbial population
Serves as a storehouse of elements essential for plants

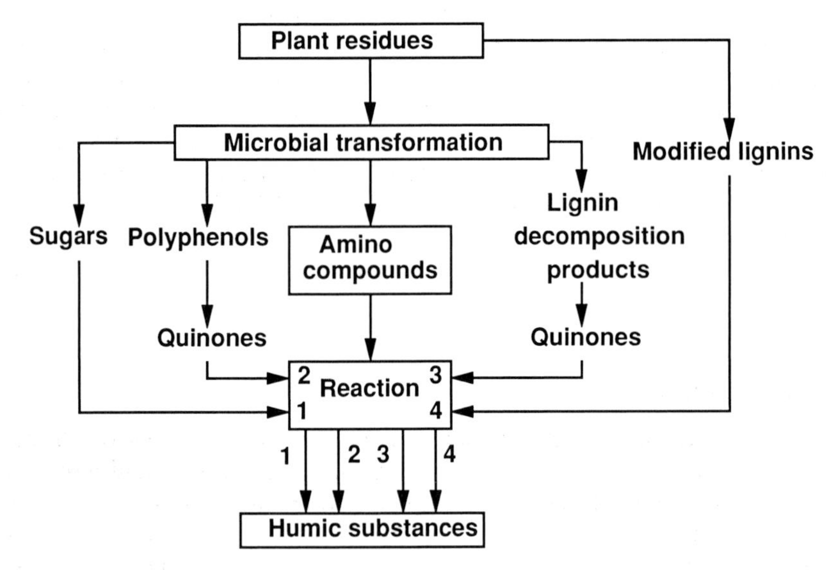

Figure 1.2 Proposed major pathways of humus formation. After F.J. Stevenson (1982).

plant and animal remains, and four major pathways are represented in Figure 1.2.

Pathway 1: reducing sugars and amino acids, formed as by-products of microbial metabolism, undergo non-enzymatic polymerisation to form brown nitrogenous polymers.

Pathway 2: polyphenols are synthesised from non-lignin sources (e.g. cellulose), and are enzymatically oxidised to quinones which polymerise in the presence or absence of amino compounds to form humus.

Pathway 3: phenolic aldehydes and acids, released from lignin during decomposition by microorganisms, undergo enzymatic conversion to quinones which polymerise as in Pathway 2 to form humus.

Pathway 4: lignin is incompletely utilised by microorganisms and the residues undergo demethylation, oxidation and condensation with N compounds such as proteins to form humus.

For many years the lignin theory of humus (Pathway 4) predominated, but the polyphenol theory (Pathways 2 and 3) is now favoured. However, a unifying theory of humus formation for all soils has not been forthcoming, and all the above pathways may be involved. It should be added that the formation of humus during composting follows similar principles, but allows greater control to be exerted over the environmental conditions.

Two fundamentally different approaches have been used to study soil organic matter. The classical approach is to use chemical extraction and analysis of soil organic matter to produce fractions such as humic acid (soluble in alkali, precipitated in acid) and fulvic acid (soluble in alkali, not precipitated in acid). However, the biological significance of these fractions is uncertain. The second approach is more empirical and attempts to classify fractions according to their rates of decompositon. The simplest approach is to consider organic matter as a mixture of labile (decomposable) and resistant substrates for organisms. This topic is considered further in chapter 4.

The overall composition of soil organic matter can be summarised as shown in Figure 1.3.

1.2.3 Soil structure

The arrangement of sand, silt, clay and organic matter in soil gives the soil structure. Inter-particle forces hold clay particles together and also bind them to sand and silt surfaces. Organic matter and hydrated Fe and Al oxides assist in stabilising these groups of particles, termed aggregates. Figure 1.1 shows a schematic arrangement for soil as seen at two levels of magnification.

The size and arrangement of the solids determine the size and shape of the spaces (pores) between them. The pores in a soil determine the air and water relationships of that soil, and therefore affect biological activity.

1.2.4 Soil moisture

The sizes of the solids and spaces have a profound effect on the soil water status. Pores, which are joined to one another by narrow channels or necks between particles, can be grouped in order of decreasing size into transmission pores, storage pores, and residual pores (Figure 1.1). The

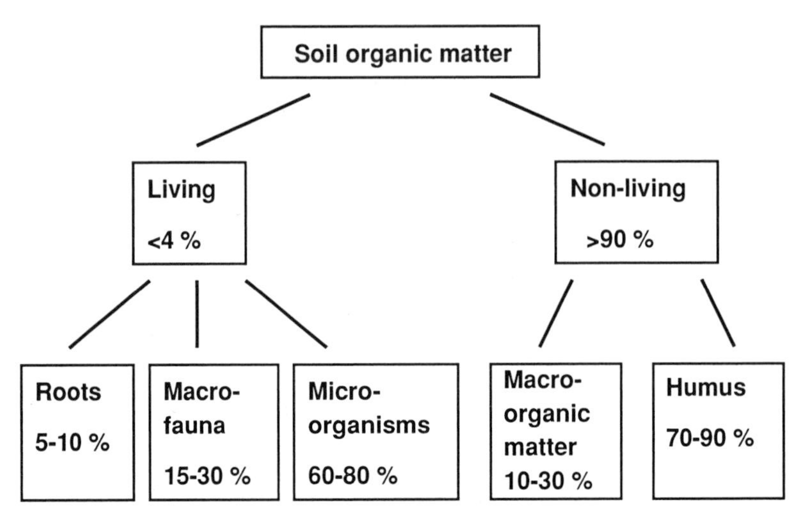

Figure 1.3 Composition of soil organic matter by weight (excluding litter). After B.K.G. Theng *et al.* (1989).

pores form a continuous three-dimensional network throughout the soil. The 'potential' of the soil water (a reflection of the work needed to remove unit quantity, and hence its availability) is of more direct importance for soil organisms than is the absolute water content.

The soil matrix is clearly very complex and gives rise to forces associated with the interfaces between water and solids, and between water and air. These forces decrease the potential of the soil water and this component of potential is called the matric potential. As the soil dries the larger pores (strictly the pores with the largest necks) are emptied first (Figure 1.4). If it is assumed that soil pores act as capillary tubes, then the pore size can be related to the matric potential of the water contained within that pore.

Field capacity is the water content of the soil draining from saturation when the loss due to gravity has become negligible. It corresponds to a matric potential of about −10 kPa (equivalent to −0.1 bar or −100 cm water column). Permanent wilting point is the water content at which plants permanently wilt, and is taken as the water content corresponding to a matric potential of −1500 kPa. These criteria have been used to classify the different types of pores found in soils according to size and function. Transmission pores (>50 μm) are drained at field capacity, but allow root penetration. Storage pores (50–0.5 μm) hold water against drainage between the limits of field capacity and permanent wilting point, this water being available for transpiration. Residual pores (<0.5 μm) hold water that is unavailable to plants, but which may be evaporated from the soil surface.

The values of water potential at both field capacity and permanent

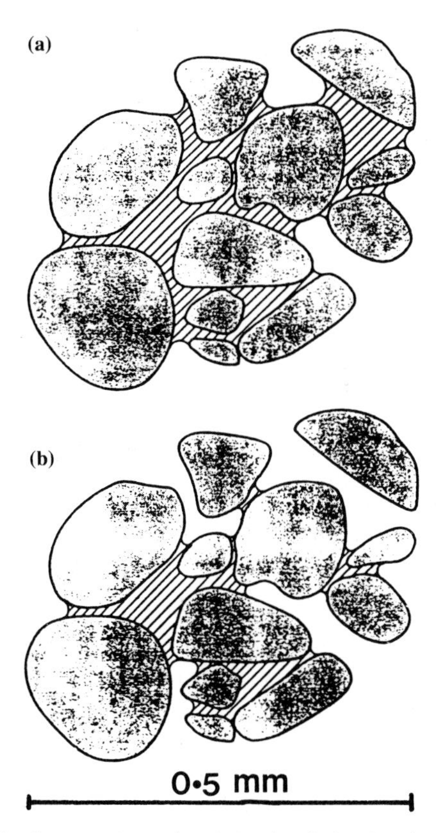

Figure 1.4 A schematic diagram of a section through soil showing the distribution of water at (a) −10 kPa and (b) −20 kPa. From D.M. Griffin (1972) *Ecology of Soil Fungi*, Chapman and Hall, London.

wilting point are to some extent arbitrary, and do not have any precise biological significance. However, well-drained soils are rarely wetter than field capacity, and surface soils may be much drier than permanent wilting point even though plants show no signs of stress, because they are supplied with water from deeper in the soil. At matric potentials less than −1500 kPa movement of water as vapour becomes increasingly more important.

Water moves in soil from areas of high water potential to areas of low water potential. Differences in water potential between the root and soil, and the resistance to movement of water determine the ability of a root to take up water for transpiration. The resistance to flow depends upon the conductivity of the soil to water (the hydraulic conductivity), which varies greatly with soil water content. The hydraulic conductivity of the soil has little direct influence on microorganisms, in contrast to plants. However, hydraulic conductivity decreases rapidly as water content decreases, and this can affect movement of organisms as they are carried along with the soil water, and slows down the redistribution of water in dry soils.

The potential of the soil water is also decreased by the presence of solutes, which are necessary for growth, and this gives rise to an osmotic component of potential. Furthermore, as soils dry, the pathways through which solutes can diffuse from areas of high concentration to low concentration also decrease. Therefore, effects of moisture stress on microorganisms in soil are confounded by several factors.

Bacteria and fungi are affected similarly by decreases in soil water potential. Growth is generally restricted to values greater than between −6000 and −800 kPa. At water potentials below −14 500 kPa *Aspergillus* sp. and *Penicillium* sp. predominate. There is no evidence of a specific xerophytic population in arid zones; fungal species capable of growth in laboratory media at −30 000 kPa are as prevalent in soils from an English pasture and an Australian rain forest as in desert soils. The apparent dominance of streptomycetes in many dry soils, as determined by plate counts, is probably due to the resistance of their spores to desiccation rather than to hyphal activity at low water potentials.

Clays offer some degree of protection to microorganisms against desiccation. The surfaces of these minerals retain water which, together with the coordination water from exchangeable cations, may assist microbes to survive dry conditions. However, this water is unlikely to be available to microorganisms and the surface tension forces in such films of water may prevent microbial movement. It has been suggested that the higher affinity of montmorillonite than cells for water causes rapid drying of rhizobia during desiccation, which increases the tolerance of these bacteria to desiccation. Overall, the mechanisms by which surfaces influence microbial survival and activity in soils are poorly understood.

1.2.5 Soil atmosphere

Biological responses to changes in soil water content may not be directly related to water but may be due to associated changes in the soil atmosphere. The soil atmosphere differs from the free atmosphere because plant roots and organisms living in soil remove O_2 and produce CO_2. This causes diffusion of O_2 into the soil and CO_2 out of the soil. In general, however, soil air has a higher concentration of CO_2 and lower concentration of O_2 than the free atmosphere.

The rate of respiration in soil depends upon soil moisture content, temperature and the amount of readily decomposable organic matter. Measurements made on a soil at Rothamsted Experimental Station indicated daily rates of O_2 consumption of 0.7–24.0 g m^{-2} and rates of CO_2 production of 1.2–35.0 g m^{-2}. Soil contains enough O_2 to support respiration for no more than a few days if the supply of O_2 is restricted.

Oxygen (O_2) is unlikely to limit biological activity at low matric potentials because most pores are filled with air, in which the diffusion rate of O_2 is 300 000 times as great as that in water (for O_2 the solubility and diffusion coefficient are, respectively, 0.03 and 0.0001 times lower in water than in air). However, at higher water potentials many of the pores are filled with water, and O_2 may become limiting. In a well-structured soil the individual particles are aggregated into crumbs containing very small pores, therefore the crumbs may remain saturated with water over a wide range of water potentials. Larger pores exist between crumbs and they will be empty of water at relatively high matric potentials.

If microorganisms are evenly distributed within a soil crumb their respiratory activity, coupled with the slow diffusion of O_2 through the crumb, can give rise to anaerobic zones at the centre of water-saturated crumbs of diameter greater than about 6 mm. These zones represent an important microsite environment within soil.

The diffusion of volatile compounds such as ethylene, a plant growth hormone, will also be affected by the soil water content, but little is known of this.

1.2.6 Temperature

Organisms are often exposed to a wide range of fluctuating temperatures in soil. Soil temperature depends upon atmospheric temperature and inputs and losses of radiation. It is also influenced by the presence or absence of vegetation, the soil water content and the depth within the soil. Temperatures vary between day and night (diurnal variation) and throughout the year. The temperature of the soil surface may reach 60°C or may be below 0°C. Diurnal fluctuations of up to 35°C can occur at the soil surface. With increasing depth, diurnal temperature fluctuations are reduced. The mean fluctuation in summer in subsurface soils in cool temperate climates may be only 10°C and in subtropical and tropical regions may be 30°C.

Microorganisms can be categorised according to the range of temperatures at which they grow. Psychrophiles grow at 0°C and have an optimum temperature at or below 20°C. Thermophiles have a maximum temperature for growth of greater than 50°C and a minimum greater than 20°C. Mesophiles have temperature optima between these two extremes. The ecological significance of these different responses to temperature is not clear. For example, thermophilic bacteria and fungi are found in temperate soils which appear more suited to psychrophilic organisms.

The ability of microorganisms to grow at low temperatures may depend upon membrane permeability and associated control of solute transport into the cell. Psychrophilic bacteria appear to synthesise increased quantities of enzymes at low temperatures to compensate for the reduction in the rates of metabolic processes at low temperature. It is often assumed

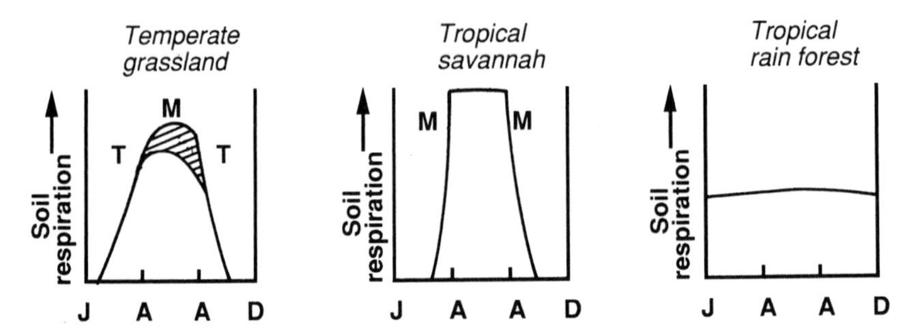

Figure 1.5 Seasonal soil respiration rates for three ecological zones, showing the main climatic limitations (T, temperature limited; M, moisture limited). After M.J. Swift, O.W. Head and J.M. Anderson (1979) *Decomposition in Terrestrial Ecosystems*, Blackwell Scientific Publications, Oxford.

that soil microbial activity is negligible at temperatures less than 5°C, although there are reports of denitrification occurring in soils at such low temperatures.

One of the few consistent differences between soils in tropical environments and soils in temperate environments is that the mean annual temperature in the tropics is roughly 15°C higher than in temperate regions. High temperatures in soils are often associated with dry conditions, and the resulting effects on soil organisms and processes can be complex. Figure 1.5 shows the effect of climate on soil respiration in three contrasting ecological zones. In temperate grassland soils, microbial activity is limited by low temperatures during winter, whereas the major limitation to soil organisms in tropical savannah soils during the same period is drought. In tropical rain forests neither temperature nor soil moisture limit microbial activity throughout the year, leading to relatively constant and rapid rates of organic matter turnover in these soils.

The effects of freezing and thawing on soil organisms are generally less pronounced than drying and re-wetting. Fluctuations in soil temperature and moisture may have a more profound effect on soil microbial processes than constant extreme conditions.

1.2.7 Summary

This chapter has provided an introduction to the main physical and chemical features of soil as a habitat for soil organisms. Further information can be found in the references and further reading provided at the end of the book.

2 Life in the soil

2.1 Introduction

The harvesting by plants of energy from sunlight, and carbon and nitrogen from the atmosphere (by photosynthesis and nitrogen fixation, respectively), and their conversion into organic compounds, together with the mining of mineral nutrients from deep within the subsoil provide the essential requirements for life in the soil.

The remains of plant material (often after being consumed by animals) and the remains of animals following death are decomposed by microorganisms. These processes of synthesis and decomposition form part of the cycle of life and death (Figure 2.1). It is worth noting at this point that many of the processes taking place in soil can be considered as components of larger natural cycles.

2.2 Microbial biomass

The fixation of CO_2 during photosynthesis provides in the form of organic compounds the carbon and energy for life in the soil. However the net primary production (the rate of production of plant biomass after allowing for losses due to respiration) varies in different ecosystems (Table 2.1). Annual production is greatest in tropical rain forests (up to 3500 g m^{-2}) and decreases as photosynthesis is reduced by extremes of temperature and moisture.

These differences in net primary production lead to differences in total organic C in soils (Table 2.2) and in soil microbial biomass. Comparison between Table 2.1 and Table 2.2 shows that although the tropical rain forest has the highest primary production, the soil that supports this vegetation contains a relatively low amount of organic matter. This is due to the rapid decomposition of plant litter in this ecosystem. Soil under temperate permanent grassland contains more organic matter due to the dense root mass in grassland soils and the slower rate of decomposition.

Soil microbial biomass has been formally defined as the living part of the soil organic matter, excluding plant roots and soil animals larger than approximately 5×10^3 μm^3. This definition is arbitrary, and based on estimates based upon direct microscopical counting. Values for microbial biomass in the soils in Table 2.2 reflect the organic matter content, an

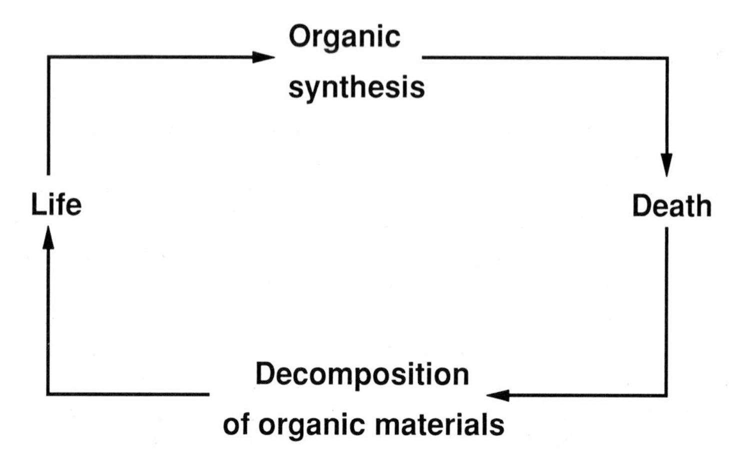

Figure 2.1 In the cycle of nature processes of decomposition are also processes of synthesis.

Table 2.1 Net primary production in different ecosystems. After R.H. Whittaker and C.E. Likens (1975) in *Primary Productivity of the Biosphere*, eds. H. Leith and R.H. Whittaker, Springer Verlag, pp. 305–328

Ecosystem	Normal range ($g\ m^{-2}$ per year)	Mean ($g\ m^{-2}$ per year)
Tropical rain forest	1000–3500	2200
Temperate deciduous forest	600–2500	1200
Savanna	200–2000	900
Temperate grassland	200–1500	600
Tundra and alpine	10–400	140
Desert and semidesert scrub	10–250	90

Table 2.2 Organic matter and microbial biomass contents (expressed as carbon) for soils from different ecosystems. After D.S. Jenkinson and J.N. Ladd (1981) Microbial biomass in soil: measurement and turnover. In *Soil Biochemistry*, Volume 5, eds. E.A. Paul and J.N. Ladd, Marcel Dekker, New York, pp. 415–471

Ecosystem	Depth (cm)	Organic C ($g\ m^{-2}$)	Biomass C ($g\ m^{-2}$)	Biomass C as % organic C
Tropical rain forest	0–15	1900	76	4.0
Temperate grassland	0–23	7000	224	3.2

indication of the amount of substrate available for maintenance and growth of soil microorganisms. The values for microbial biomass (in terms of C) form a fairly constant proportion of the total organic C, ranging from 2–4% (Table 2.2).

2.3 The soil inhabitants

The concept of soil microbial biomass has been useful in advancing our understanding of organic matter dynamics. However, in order to understand the functioning of the soil population in the many roles in which it is involved it is necessary to obtain information on the different groups of soil inhabitants. Such knowledge is also of vital importance in the quest to conserve soil biodiversity (see chapter 6). The remainder of this chapter therefore introduces some of the major players in life in the soil.

There have been only a small number of studies which have quantified the major groups of organisms in particular soils, and only a few of these have considered the interactions which might occur between these different groups. This paucity of information is partly due to the lack of reliable techniques for quantifying the numbers and sizes of the different organisms in soil, and also because of the labour-intensive nature of the existing techniques. Some studies have identified individual species present in soil, but have not related this information to the activity of these organisms.

2.3.1 Functional groups of organisms

The role of particular organisms can be deduced in a variety of ways; this allows the organisms to be arranged into functional groups. Specific groups can be added to sterile soil (this system is termed a microcosm) and the effect of this addition on a process such as N mineralisation can be measured. Conversely, selective biocides may be used to eliminate specific groups of organisms such as nematodes (by using nematicides) and the effect of this removal measured. Alternatively, the dynamics of major groups can be followed in soil and changes in particular groups correlated with particular processes. Organisms such as nematodes may also be grouped according to morphology (e.g. mouthparts) and other biological characteristics.

Organisms may be classified according to their body length or width (Table 2.3). The use of litter bags with different mesh sizes to exclude particular groups of organisms allows the role of these organisms in decomposition of litter to be deduced. Such experiments suggest that the microfauna are rarely involved in litter comminution, the mesofauna do attack plant litter, but their overall contribution is small (termites are the exception to this) and the macrofauna have the major effect on litter decomposition. The role of earthworms and termites in soil formation and development is discussed further in chapter 4.

Two sets of data are now presented to show the detailed composition of the soil population in two contrasting ecosystems. Table 2.4 shows the

Table 2.3 Classification of soil organisms according to body width. After M.J. Swift, O.W. Heal and J.M. Anderson (1979) *Decomposition in Terrestrial Ecosystems*, Blackwell Scientific Publications, Oxford

Grouping	Body width	Organisms
Microflora	<10 μm	Bacteria, fungi
Microfauna	10–100 μm	Protozoa, nematodes
Mesofauna	0.1–2.0 mm	Collembola, acari, enchytraeids, termites
Macrofauna	2–20 mm	Millipedes, isopods, insects, molluscs, earthworms

Table 2.4 Biomass estimates for functional groups of organisms in a woodland soil (Meathop, UK). After J.E. Satchell (1971) in *Productivity of Forest Ecosystems*, Unesco, Paris, pp. 619–629, and T.R.G. Gray *et al.* (1974) *Révue d'Ecologie et de Biologie due Sol* **11**, 15–26

Functional group	Biomass ($g\ m^{-2}$)
Bacteria	3.7
Fungi	45.4
Protozoa	0.1
Nematodes	0.2
Enchytraeids	0.4
Earthworms	1.2
Molluscs	0.5
Acari	0.1
Collembola	0.2
Diptera	0.3
Other arthropods	0.6
(Annual litter production	764.0)

estimated biomass of groups of organisms in a woodland soil in the UK. This is a mixed deciduous forest in the north west of England with an acid soil. Total annual net primary production for the forest has been estimated to be 1308 $g\ m^{-2}$. This system is not at steady state; 46% of above-ground production is added to the standing crop, the remainder (546 $g\ m^{-2}$) enters the soil as litter or leachates. This acid soil is dominated by fungi (microflora) with a biomass of 45.4 $g\ m^{-2}$ compared to the total biomass of the soil and litter fauna of 3.6 $g\ m^{-2}$.

Data shown in Table 2.5 are for a prairie soil in the USA. In this case it has been assumed that only 10% of the total fungal hyphae are active. The microflora also dominate this soil, but the main functional group in this near-neutral soil is the bacteria. The subdivision of groups such as nematodes reflects an increased understanding of the roles of these organisms. Ciliates, earthworms, termites and insect larvae were not major components of this particular ecosystem.

Table 2.5 Biomass estimates for functional groups of organisms in a shortgrass prairie soil (Colorado, USA). After H.W. Hunt *et al.* (1987) *Biology and Fertility of Soils* **3**, 57–68

Functional group	Biomass ($g\ m^{-2}$)
Bacteria	60.80
Saprophytic fungi	1.26
VA mycorrhizas	0.14
Amoebae	0.76
Flagellates	0.03
Phytophagous nematodes	0.06
Fungivorous nematodes	0.08
Bacterivorous nematodes	1.16
Omnivorous nematodes	0.13
Predaceous namatodes	0.22
Fungivorous mites	0.61
Nematophagous mites	0.03
Predaceous mites	0.03
Collembola	0.01

Studies have often concentrated on bacteria and fungi, which form the largest groups in terms of biomass in the soil, without considering the higher trophic levels. However, groups of organisms such as mites, although a small component, may be important in determining rates of nutrient cycling and the stability of the soil ecosystem.

The main functional groups in soil are now considered. Plants, or more strictly plant roots, are one of the major inhabitants of soil and they are also included here.

2.3.2 Viruses

Viruses are a group of non-cellular infectious agents. They are obligate parasites of cellular organisms and may be more correctly considered as complex molecules of protein and nucleic acid, rather than organisms. They have no metabolism of their own, they vary in shape, and the virus particle (a virion) ranges in diameter from 20–250 nm (0.02–0.25 μm).

A range of plant, insect and human viruses is found in soil, and they often persist for several years. For example, plant viruses survive in soil for up to 9 years in the absence of their host and the nuclear polyhydrosis virus of the cabbage looper (*Trichoplusia ni*), although surviving only one month on the leaves, persists for up to 4 years in soil.

Human viruses such as picornaviruses are introduced into soil in the effluent or sludge from sewage (waste water) treatment works. Human pathogens including helminthic parasites and human enteric viruses are neither completely removed nor inactivated by conventional sewage treatment methods such as trickling filter and activated sludge techniques.

Little is known about the survival of human viruses in soil, but poliovirus can survive in soil for 3–4 weeks. Soil conditions such as moisture and the types of adsorbing surfaces, e.g. clay minerals, are likely to be important in affecting the survival and movement of viruses in soil.

2.3.3 Bacteria and actinomycetes

Bacteria and actinomycetes are the smallest organisms in soil (unless viruses are considered as organisms). They are prokaryotes: organisms with a relatively simple cell structure when viewed under the transmission electron microscope, with no internal organelles. All the other groups are eukaryotes: organisms with a relatively complex internal cell structure, with nuclear membrane, mitochondria and, in some organisms, chloroplasts.

Although early workers considered actinomycetes, with their characteristic mycelium, to be fungi, later studies showed them to be bacteria. Bacteria can be grouped according to their reaction with Gram's stain. This depends upon cell wall components; those organisms which retain the stain are termed Gram-positive, and those that do not retain the stain are termed Gram-negative.

Bacteria are generally rod or cocci shaped and up to several micrometres in length and approximately 10^{-12} g in weight and 1 μm^3 in volume. Accurate estimates of numbers of bacteria and actinomycetes in soils are difficult to obtain. Numbers depend upon the methods used; dilution plate techniques yield estimates which are up to 500 times lower than those obtained using direct microscopy. There is clearly a large number of bacteria present in soil which are viable, but which cannot be grown in conventional media. It is salutary to realise that most of our understanding of bacteria in soil is based upon those organisms which can be cultured in the laboratory.

The numbers of bacteria in 1 gram of soil vary from 10^6 to 10^9. Numbers also fluctuate with time; Figure 2.2 shows the variation in plate counts and direct counts for bacteria in samples of a Rothamsted soil taken from the field every 2 h for 24 h.

Bacteria are not uniformly distributed throughout the soil profile, nor throughout a single soil horizon, but it is generally observed that their distribution follows that of the soil organic matter. Table 2.6 shows the decrease in numbers of bacteria in a soil profile. The numbers actinomycetes also decrease with soil depth, but they increase in proportion to the other bacteria from 10–65%.

Bacteria are located in small colonies (Figure 2.3a, c) often associated with sources of organic substrates (for example, plant roots).

Table 2.7 shows estimated numbers of bacteria and actinomycetes in some tropical soils. These data are similar to those reported for temperate

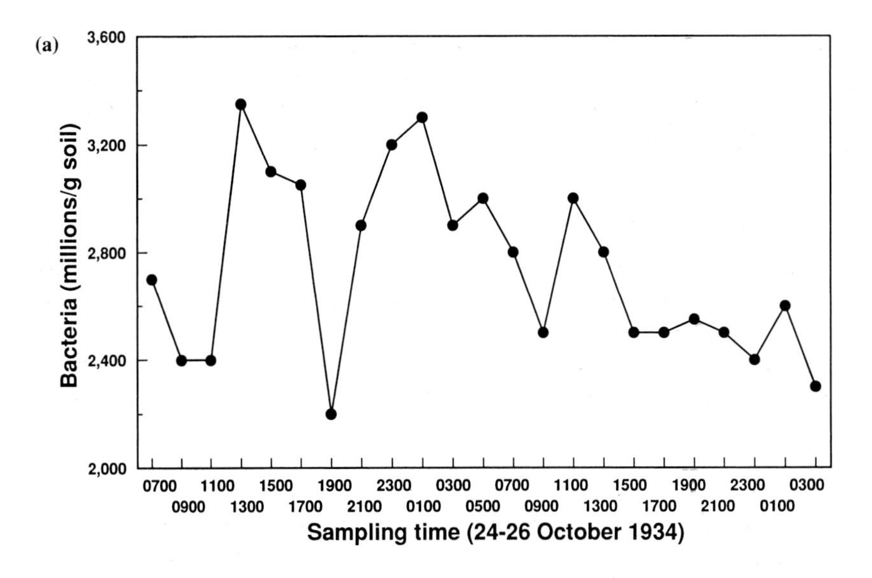

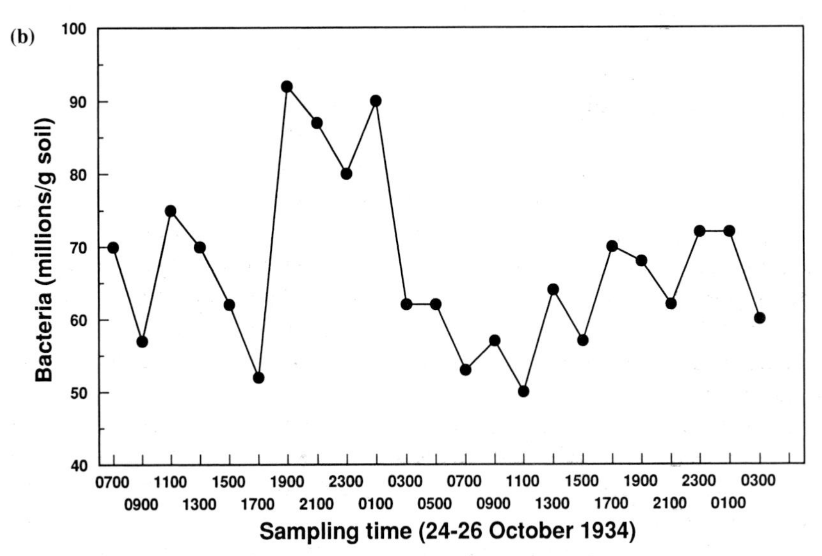

Figure 2.2 Numbers of bacteria as determined by (a) direct counts, (b) dilution plate counts in a field soil at Rothamsted, England over a 44 h period. After C.B. Taylor (1936) *Proceedings of the Royal Society of London B* **119**, 269–295.

soils. It should be remembered that although on a numerical basis the bacteria may outnumber the actinomycetes by a factor of 10 (and the fungi by a factor of 100), the bacterial biomass in many cultivated soils is about equal to that of the actinomycetes and only about half that of the fungi.

Table 2.6 The distribution of aerobic bacteria in a soil profile. After A. Starc (1942) *Archives of Microbiology* **12**, 329–352

Horizon	Depth (cm)	Number (10^6 per gram)
A1	3–8	7.8
A2	20–25	1.8
A2/B1	35–40	0.47
B1	65–75	0.01
B2	135–145	0.001

Table 2.7 Estimates of numbers of bacteria and actinomycetes in some soils of the humid tropics (after A. Ayanaba and F.E. Sanders, 1981)

Method	Number (10^6 per gram)	Country
Bacteria		
Direct	2000–3300	Kenya
Plating	14–340	Kenya
Plating	1–11	Malaysia
Plating	1–118	Nigeria
Actinomycetes		
Plating	2–37	Kenya
Plating	710	Nigeria
Plating	10–79	Nigeria

The biochemical diversity of bacteria and actinomycetes as a group has a major impact on soil. There are few natural or xenobiotic compounds which cannot be broken down by these organisms. For example, strains of *Pseudomonas* sp. can grow on industrial pollutants such as 3 chlorobenzene, and many actinomycetes can use chitin as a source of energy, carbon and nitrogen. Bacteria have evolved many unusual biochemical characteristics, for example *Thiobacillus ferrooxidans* obtains energy from oxidation of reduced sulphur compounds and ferrous ions, and has an optimum pH for growth of 2. *Clostridium* spp. are able to grow in the absence of oxygen and can obtain N by the reduction of atmospheric N_2 gas. *Rhizobium* spp. form N_2-fixing nodules on the roots of leguminous plants.

The activities of bacteria in soil have important implications for the cycling of nutrients such as N and S which are derived from litter. The processes of denitrification and sulphate reduction involve a variety of facultative and obligate anaerobic bacteria, respectively. Ammonium

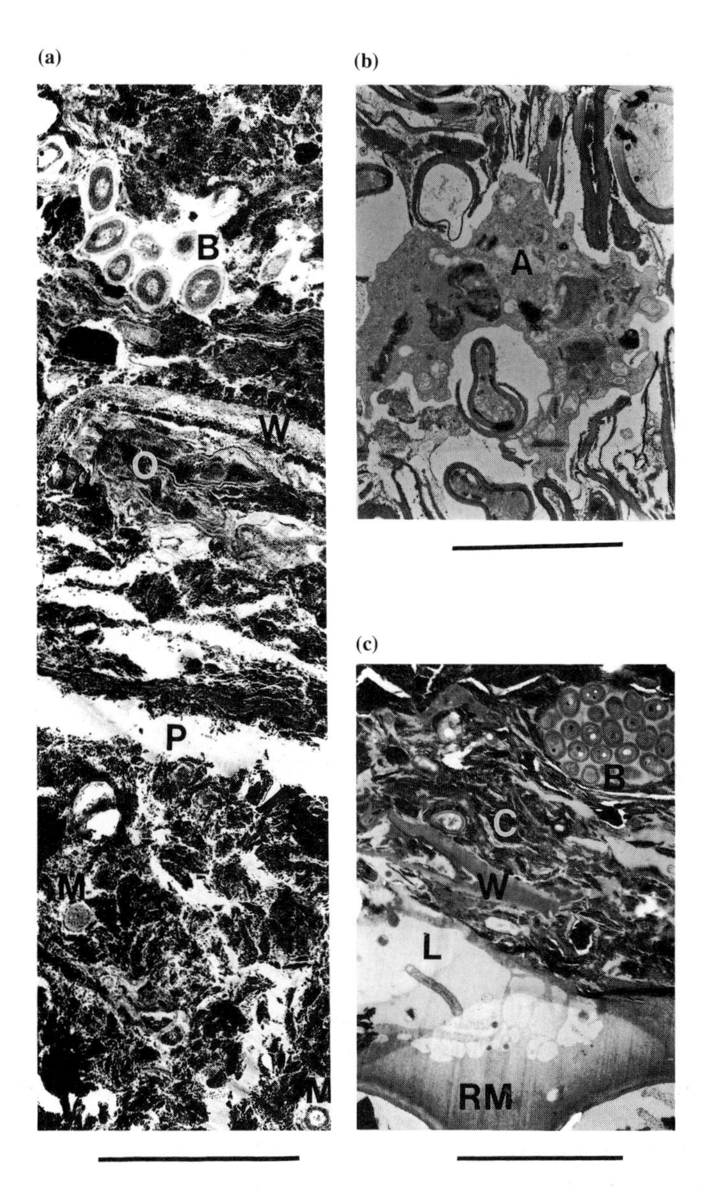

Figure 2.3 Electron micrographs of soil: (a) bacteria (B) associated with organic matter which still contains carbohydrate (e.g. cell walls, W); other highly lignified and convoluted organic matter (O) does not support bacteria, but there are numerous microorganisms (M) scattered throughout the clay; P is a pore about 1 μm in diameter; (b) an amoeba (A) in an organic-rich surface soil; (c) root surface mucilage (RM) which has been partially lysed (L) by soil bacteria; the mucilage holds cell remnants (W), clay particles (C) and a colony of bacteria (B) onto the root surface. Courtesy of Dr R.C. Foster, CSIRO, Australia.

oxidation (nitrification) and sulphur oxidation involve only a few genera and species of aerobic autotrophic bacteria. These processes are described in more detail in chapter 3.

Although there have been relatively few detailed studies on the microbiology of soils in the tropics, there is no evidence of any great differences from soils in temperate regions which have received greater attention. For example, nitrifying bacteria isolated from a Kenya highland soil appear morphologically and physiologically similar to those isolated from temperate soils.

Of the 190 genera of bacteria listed in the 7th Edition of Bergey's *Manual of Determinative Bacteriology*, 97 genera (51%) contained species that may be considered as soil bacteria. Corynebacteria, for example *Arthrobacter* sp., form at least half of the total bacterial colonies which grow on dilution plates. Spore-forming bacilli and actinomycetes are also common. These bacteria, which bring about many of the major biochemical transformations in soil, probably constitute less than 10% of the total bacteria population; the function of the remaining population is not known.

Many soil coryneform bacteria, including *Arthrobacter* sp., are major catabolisers of heterocyclic compounds and hydrocarbons, and a strain capable of degrading lignin has been isolated from decaying peanut hull. Actinomycetes may play an important role in the formation of humic acids from litter in soils.

Some bacteria and actinomycetes form symbiotic relationships with plant roots: *Rhizobium* spp. form nitrogen fixing nodules with leguminous plants and *Frankia* spp. form actinorhizal nodules on roots of a wide range of perennial woody trees and shrubs (sections 2.4.3 and 2.4.4).

Some of the metabolites produced by actinomycetes are of particular interest. Many *Streptomyces* spp. isolated from soil are capable of producing antibiotics under laboratory conditions which are useful for medical purposes. The discovery of the antibiotic streptomycin, which is active against a wide range of bacteria including *Mycobacterium tuberculosis,* was founded on work carried out on a soil isolate of *Streptomyces griseus. Streptomyces* spp. also produce an earthy odour which is reminiscent of newly turned soil, and this has been linked to a secondary metabolite called geosmin.

Soils may also contain some bacteria which are pathogenic to humans. For example, *Clostridium* spp. are common in soil and animal faeces. They are anaerobic organisms which produce spores, and some species such as *C. tetani* and *C. perfringens* can enter wounds and cause the diseases tetanus and gas gangrene, respectively. These diseases may be lethal; prophylactic immunisation is used to prevent tetanus.

Bacillus anthracis is the only species of Bacillus, a common soil genus, which is pathogenic to humans. Like all bacilli it produces spores which, in this case, cause the disease anthrax, and which may persist in soil for many years. In the 1940s the island of Gruinard off the north coast of Scotland was infected with spores of *B. anthracis* as part of a biological weapons test. The persistence of these spores in the soil meant that the island was uninhabitable until 1987 when the spores were eradicated by treating the soil with formaldehyde.

An early attempt at a functional classification for soil bacteria by Winogradsky in the early part of this century drew a distinction between those bacteria which exist mainly in a resting state with brief periods of activity when substrates are available (termed zymogenous), and those which are continually active, but at a low level (termed autochthonous).

2.3.4 Fungi

Fungi produce filamentous mycelia composed of individual hyphae, 0.5–10 μm in diameter, which may or may not be subdivided by crosswalls (septa). Fungi are less biochemically versatile than bacteria, for example they are all aerobic heterotrophs (see chapter 3), but they are more variable in their morphology, for example in the range of fruiting bodies they produce. From 1 to 75% of total hyphae are active depending on soil conditions, and the hyphae are often irregular in shape and size due to close contact with soil particles.

There are many lists of genera of fungi isolated from soil, but most of this information is obtained from the dilution plate technique. However, many soil fungi, such as the larger Ascomycetes and Basidiomycetes, do not grow or sporulate readily on soil dilution plates. Over 690 species of fungi from 170 genera have been isolated from soil, with a small number of genera, such as *Penicillium* and *Aspergillus,* accounting for more than half of the species.

Fungi produce a range of specialised structures in soil. Thickened strands of mycelia, termed rhizomorphs, are often conspicuous in the litter layers of woodland soils. Thick-walled structures such as chlamydospores (vegetative resting spores), vesicles containing oil globules, and sclerotia which consist of closely interwoven hyphae 20–30 cm in length, are widespread. Fungi also produce a variety of spores which appear to have only limited survival ability. A wide range of fruiting bodies are formed which vary in size and complexity. Most of the larger fruiting bodies produced by Ascomycetes and Basidiomycetes occur at the soil surface, for example those of agarics found in woodland and pastures. Some fungi are also important plant pathogens (section 2.4.2), for example *Gaeumanno-myces graminis*, the causative agent of take-all disease in wheat, which overwinters in soil on debris from the previous crop ready to infect the following crop.

Other fungi form a symbiosis with plant roots, termed a mycorrhiza, which is important in improving the supply of phosphorus to the plant (section 2.4.6).

2.3.5 Cyanobacteria and algae

Cyanobacteria, formerly known as blue-green algae, and green algae (chlorophyceae) are photosynthetic organisms ubiquitous in soils. Data for a range of virgin soils and cultivated chernozems in the former USSR indicate values for cyanobacterial and algal biomass of 0.7–54.6 g m^{-2}. They are usually most abundant close to the soil surface, but may also be found at depths down to 15–20 cm below the surface, where there is insufficient light for photosynthesis. The contribution of these organisms to net primary production is not clear.

2.3.6 Protozoa

Soil protozoa consist of flagellates (mean diameter 5 μm), for example *Bodo* sp., amoebae or rhizopods (mean diameter 10 μm), for example *Euglypha* sp., and ciliates (mean diameter 20 μm), for example *Colpoda* sp. Amoebae (Figure 2.3b) either consist of a mass of protoplasm surrounded only by a flexible pellicle (naked amoebae), or possess a rigid shell made of chitin or silica with fine protoplasmic strands emerging from the end (testaceans). Flagellates have one or more flagella attached to one end of their bodies which allow movement, whereas ciliates use cilia which are often arranged in bands.

There is little field information on protozoa, partly due to the difficulty in studying these microorganisms. They lack a proper cell wall and changes in pH and salt concentration, or mechanical damage, can cause the cells to burst. Furthermore, they are small and variable in shape, and because their numbers are at least an order of magnitude less than for bacteria, direct counting is impossible. The biomass of protozoa in an arable soil can be almost equal to that of the earthworm population, but protozoa have a much shorter turnover time. A Rothamsted soil contained 70 500 flagellates per gram, 1400 amoeba per gram, and 377 ciliates per gram. Although data often indicate that flagellates and amoebae are more numerous than ciliates, when the mass of the individual organisms is taken into account (0.2–28.0 ng for flagellates, 0.8–6.0 ng for amoebae and 1.5–750 ng for ciliates) then the ciliate biomass may equal or exceed that of the other two groups.

Protozoa are major consumers of bacteria, and they are able to survive adverse conditions by cyst formation. They have been recovered from air-dry soil after being stored for 49 years. Naked amoebae appear more suited to soil because their sliding motion on surfaces enables them to feed on soil particles and roots, and their flexible cells are adapted to feeding within

thin water films around soil particles. Many flagellates are saprozoic, feeding on dissolved organic nutrients. Ciliates, being larger organisms, are probably restricted to feeding during periods of high water content.

A neglected and unusual group of soil organisms with some characteristics in common with protozoa are the myxomycetes. They are found in a wide range of soils from Albanian woodland to the Sonoran desert and possess a combination of plant, microbe and animal features. The germination of spores produces myxamoeba or flagellate cells which give rise to swarmers and swarm cells (10–20 μm long). Myxomycete biomass has been estimated as 0.006–0.064 g m^{-2}. Little is known of the ecology of these organisms although the dung of herbivorous and omnivorous mammals provides substrates for their growth. They consume other microorganisms and are eaten by a range of organisms.

2.3.7 Nematodes

Nematodes are a diverse group of roundworms widely distributed in soils, and when adult may be 0.5–1.5 mm long and 10–30 μm diameter. They depend upon a thin film of water around soil particles for movement and completion of their life cycle; growth is better in coarse-textured soils than in fine-textured soils. However, their microscopic size, clumped distribution and variety of feeding habits present problems when investigating their role in soil. Their distribution in soil generally follows that of organic matter.

Nematodes can be subdivided into functional groups. Some nematodes, for example *Meloidogyne* sp., are plant parasites (section 2.4.7), but the majority (70%) do not feed directly on plant roots but consume other soil organisms. However, nematodes may indirectly affect plants by feeding on fungal pathogens or increasing mineralisation of nutrients. For example, the fungivorous nematode *Aphelenchus avenae* decreases the population of the fungal pathogen *Rhizoctonia solani*. Bacterivorous nematodes can survive on bacteria such as *Agrobacterium tumefaciens* and *Erwinia carotovora*, and 30–60% of bacteria engulfed may be defecated in a viable condition. Omnivorous nematodes feed on all trophic levels from root hairs, bacteria and fungi to protozoa and other nematodes.

Predatory nematodes consume protozoa, rotifers (organisms the same size as protozoa recognised by the presence of a crown of cilia), tardigrades (soft-skinned organisms less than 1 mm long with eight legs which resemble microscopic bears), fungal spores, enchytraeids (section 2.3.8) and other nematodes. Nematodes are prey to certain fungi, for example *Arthrobotrys oligospora,* protozoa and other nematodes. In some species, stress conditions may induce sex reversals within individuals leading to a male-biased sex ratio, which may assist survival.

2.3.8 *Earthworms*

Earthworms are found in many soils throughout the world. Populations range from less than 1 to 850 per m^2 (0.5–300 g m^{-2}). Data on populations from a range of soils are given in Table 2.8. Comparisons between different soils are complicated by differences in extraction methods and in seasonal changes for a particular soil. However, there are generally fewer earthworms in acid soils (for example heathland) and bare fallow soils than in pasture soils. Numbers in arable soils are particularly variable. In the humid tropics earthworms are best represented in grasslands and are less abundant in forested and dry areas.

Earthworms are not uniformly distributed in a particular soil, for example *Lumbricus rubellus* is more frequently found beneath dung pats. Different species inhabit different depths in soil, for example *Dendrobaena octaedra* lives mainly in the surface organic horizon, whereas *Lumbricus terrestris* commonly burrows down to 1 m and sometimes to 2.5 m. Subterranean species are often paler in colour than surface feeders.

Earthworms can migrate both horizontally and vertically in soil. *Allolobophora caliginosa* introduced into new polders in The Netherlands migrated horizontally at an annual rate of 6 m. In most parts of the world earthworms show a seasonal vertical migration, probably caused by the upper soil horizons becoming unsuitable for feeding and growth. Soil temperature and moisture content seem to be the major factors determining activity. For example, *Lumbricus terrestris* and *Allolobophora caliginosa* are most active in English pastures between the months of August and December and between April and May. When the temperature is greater than 5°C *Allolobophora caliginosa* is found at the soil surface but at lower temperatures moves deeper into the soil. In the tropics most activity occurs at the start of the rainy season. In West Africa

Table 2.8 Earthworm populations in a range of soils. After C.A. Edwards and J.R. Lofty (1977)

Site	Population (number per m^2)	Biomass (g m^{-2})
Fallow soil (former USSR)	19–34	4.6–8.4
Woodland (USA)	14–142	26–280
Arable (UK)	18	1.6
Pasture (Australia)	260–640	51–152
Savannah (Ivory Coast)	18	1.7

Millsonia anomala becomes inactive when the soil moisture content is less than 7%.

Worms also exhibit diurnal patterns of activity. *Lumbricus terrestris* is most active between 6 pm and 6 am, whereas *Millsonia anomala* has two peaks of activity, one at midnight and one at 9 am. Their main role in soil is in the comminution of plant material prior to decomposition by microorganisms. The role of earthworms in the development of soils is discussed further in chapter 4.

There are many reports of an increase in the numbers of microorganisms in the earthworm gut, or in cast material, relative to the surrounding soil. The role of these microorganisms is unclear, but it seems likely that they form an essential part of the earthworm diet, enabling the animal to grow. Most studies have been on *Eisenia foetida* (the brandling or tiger worm, found worldwide in compost heaps) because of its potential use in bioconversion of organic wastes. There is no evidence of a specialised gut microflora in *Lumbricus terrestris.* The low residence time for ingested material (20 h when feeding, 12 h when burrowing) suggests that little decomposition of resistant material occurs in the gut. There may be some breakdown of cellulose and chitin due to the production of cellulase and chitinase by the earthworm rather than by the microorganisms in the gut.

E. foetida does not gain weight when feeding on mineral soil or cellulose, but does when feeding on microorganisms. Bacteria and protozoa are preferred to fungi, and protozoa may be essential for earthworm development. Pathogenic bacteria can be utilised. The mineral fraction of the soil also appears essential for growth. Maximum weight gain occurs on substrates with a low C:N ratio (15–35). The role of nematodes in earthworm nutrition is unclear. Field data for three New Zealand pasture soils showed that the presence of earthworms caused a 50% reduction in the total number of nematodes. Furthermore, *L. terrestris* feeding on cattle dung contains a large number of nematodes in the gut, but none in the casts.

Earthworm casts, rich in ammonia and partially digested organic matter, provide a good environment for microbial growth. Increased mineralisation in cast material relative to non-cast soil may be due to an indirect effect of worms in providing a suitable environment, or an additional direct effect in providing an inoculum.

Earthworms are extremely sensitive to soil disturbance; they cannot hear but are sensitive to vibrations. Darwin noted that worms emerging from their burrows in soil in pots took no notice of sounds from a piano until they were placed on top of the piano when they retreated at once back into their burrows.

Enchytraeids are small worms 0.1–5.0 cm long which feed upon microorganisms, nematodes and plant litter. Populations of approximately 200 000 per m^2 have been found in heathland soils in Denmark and in coniferous woodland in the UK. The smaller enchytraeids may be confused with soil nematodes, but they lack the buccal stylets for piercing plants and a muscular oesophagus for sucking which are found in nematodes. The precise role of these organisms, for example *Fridericia* sp., is unclear.

2.3.9 Arthropods

Arthropods are animals with exoskeletons and jointed legs. They include centipedes (Chilopoda) which are primarily carnivorous but which may also feed upon plant tissue, and millipedes (Diplopoda) (Figure 2.4) which are vegetarian, feeding on plant material in various stages of decay. Both these groups are mainly woodland species. Termites (Isoptera) are important in tropical and subtropical regions, fulfilling a role similar to that of earthworms in temperate regions. The role of termites is discussed further in chapter 4.

Beetles and wireworms (Coleoptera), midges and gnats (Diptera) and woodlice (Isopoda) are also found in soil. Woodlice, which may feed on dead organic matter (saprophagous) or on living plants (phytophagous), are important in soils which are periodically too dry for earthworms.

The most important members of the soil arthropod population are the mites (Acari) and springtails (Collembola). A mean arthropod population of 220 000 per m^2 has been recorded for an old grassland soil in Ireland. Arthropods can be subdivided into functional groups. Predatory arthropods either pursue prey, as do carabid beetles, or ambush prey, often injecting toxins or entangling the prey with silk.

Figure 2.4 A millipede (*Polyzonium germanicum*) (approximately 2 cm long) feeding on leaf litter. Courtesy of Dr Steve Hopkin, The University of Reading.

Microphytophages feed on bacteria, fungi and algae, with fungi being the most important food. Fungivores may engulf their prey, and stimulate fungal growth by grazing senescent hyphae, or may pierce fungal cell walls and drain the fluid contents leaving a ghost hypha. Detritivores scavenge dead organic material, usually after it has been conditioned by micro-organisms. This activity by millipedes, woodlice, termites and some mites leads to comminution of organic material, stimulation of microorganisms and deposition of faeces which can contribute to soil development.

Macrophytophages may be root sap feeders such as root aphids, or root grazers such as root weevils. Finally, a number of collembolans, mites and insects are omnivores, feeding on nematodes and other soft-bodied soil invertebrates. Arthropods have their greatest effect in soils dominated by fungi, for example forest soils, where millipedes, mites and collembola are dominant.

2.3.10 Molluscs

Slugs and snails (Gasteropods) are the major molluscs found in soil. Slugs are nocturnal animals, normally active from about 2 h after sunset to 2 h before sunrise. A large proportion of the population of smaller slugs shelter beneath the soil surface. Slugs are a major pest in gardens, feeding mainly on plants, and can cause extensive damage to seedling crops during damp conditions. A slug population of 140 per m^2 has been recorded in a badly damaged wheat crop. The biomass of slugs in a garden soil may be 20–45 g m^{-2}. The dusky slug *Arion subfuscus* feeds mainly on fungi and faeces of animals.

Some slugs, for example *Milax budapestensis* ingest soil and this action, together with the production of mucus by slugs and snails when crawling, may contribute to soil structure. Faeces also contain a high proportion of partially decomposed organic matter. Slugs are consumed by toads and grass snakes.

Snails have a shell which is large enough to house them and are less destructive than slugs. Thrushes eat the polymorphic land snail *Cepea nemoralis* and hedgehogs (*Erinaceus europaeus*) feed on many species of snails and slugs. Average populations of snails of 2 per m^2 have been reported.

2.3.11 Other animals

A range of large animals such as rabbits, foxes, wombats and aardvarks form burrows in soil. It has been estimated that the average biomass of deer, badgers, foxes, moles and voles in UK woodland is 0.8 g m^{-2}.

The common mole (*Talpa europaea*) is widespread throughout the UK, but is absent from Ireland. The burrowing habit of the mole results in the formation of tunnel-like runs and molehills, which may be inconvenient to farmers and gardeners. Molehills may become colonised by ants which consolidate and enlarge them. A tunnel system may extend for 400–2000 m^2 and consist of deep tunnels 5–20 cm below the surface, and surface tunnels. The fore-limbs of the mole are specialised to form shovel-shaped digging organs. Moles are active almost continuously for about 4.5 h, alternating with rest periods of about 3.5 h. They feed largely on earthworms, but also eat insect larvae, slugs and earthworm cocoons. They have few predators apart from occasional tawny owls and foxes.

2.4 Plant roots

All higher plants have roots, although the fraction of the plant's mass that is root varies widely. Nearly all roots are cylindrical, but roots and root systems exhibit a wide range of different forms. These forms are determined by plant genetic factors, by the age of the plant, and by soil conditions.

Roots growing through soil will encounter mechanical impedance; the regular spiraling motion (nutation) of the extending root tip allows the individual root to find the path of least resistance. Roots therefore tend to follow cracks and channels (e.g. those formed by earthworms) in soil.

Most roots growing in soil will bear a dense cluster of root hairs numbering 100–1000 cm^{-1} root, and 0.1–1.0 mm long. The hairs appear a few millimetres behind the root tip and increase the effective volume of soil being exploited by roots for water and nutrients. Furthermore, most root systems are mycorrhizal (section 2.4.6).

In addition to exerting direct biological effects, roots may also affect the chemical and physical properties of the soil, thereby indirectly influencing soil microorganisms.

2.4.1. The rhizosphere

At the beginning of this century it was observed that bacteria were more abundant in the soil surrounding plant roots than in soil further away from the root; this zone of soil was termed the rhizosphere. This definition has more recently been elaborated to distinguish the surface of the root (the rhizoplane) and the outer zone of the root itself (the endorhizosphere).

The rhizosphere effect can be demonstrated by removing a plant from soil, shaking off loosely adhering soil, and then estimating, using plate counts, the number of bacteria or fungi per gram of 'firmly adhering' (rhizosphere) soil. The ratio of this number to the count per gram of non-rhizosphere soil gives an R:S ratio. Such data (Table 2.9) suggest a stimulation of growth of specific groups of organisms in the rhizosphere, for example, denitrifiers with an R:S value of 1260.

Table 2.9 The numbers of different groups of microorganisms (on a soil dry weight basis) in the rhizosphere (R) of wheat (*Triticum aestivum*) and in non-rhizosphere (S) soil, and the calculated R:S ratio. After J.W. Rouatt *et al.* (1960) *Proceedings of the Soil Science Society of America* **24**, 271–273

Microorganisms	Rhizosphere (number per gram)	Non-rhizosphere (number per gram)	R:S ratio
Bacteria	120×10^7	5×10^7	24.0
Fungi	12×10^5	1×10^5	12.0
Protozoa	24×10^2	10×10^2	2.4
Ammonifiers	500×10^6	4×10^6	125.0
Denitrifiers	1260×10^5	1×10^5	1260.0

The stimulation of microbial activity around plant roots is primarily due to the leaky nature of roots which provide substrates for microbial growth. Only 5–10% of the root surface appears to be covered by microorganisms when viewed under the light microscope. This is probably due to the small number of points of direct contact between sources of inoculum (for example, organic debris) and the root surface. The distribution appears irregular, with small clumps and larger aggregates often associated with cell junctions and moribund cells, reflecting the supply of substrates from the root (Figure 2.3c). This is considered further in chapter 3.

2.4.2 *Plant root pathogens*

Roots growing through soil are exposed to infection by pathogens, particularly by fungi. Infection is more likely if the root is damaged by abrasion with soil particles or by the emergence of lateral roots. However, as with symbionts such as *Rhizobium* sp. and mycorrhizal fungi, potential pathogens must compete with the rhizosphere microflora for substrates for growth and infection sites. As only a small portion of the surface area of young roots is covered by bacteria, this affords little protection against pathogens which move to the root and infect rapidly, for example *Phytophthora cinnamomi*.

The ability of a pathogen to infect a root depends upon the size of the fungal propagules (an indication of the reserves for growth through soil), the number of propagules, the rooting intensity of the plant and the ability of the pathogen to overcome the resistance of the host to infection. Organisms with smaller propagules need to produce more of them to come near enough to a root for colonisation. Plant species with low rooting densities may avoid infections by pathogen populations likely to give many infection points on a species with a higher rooting density.

Although fungi prefer to grow along cell junctions, the ability of fungi to translocate nutrients allows them to grow across areas of low substrate supply. This may also allow some escape from antagonistic bacteria.

However, roots will normally outgrow fungi and, therefore, lateral colonisation from soil remains an important means of infection under conditions favourable for root growth.

The fungal flora of the root changes as the root ages. Many of the fungi which develop on young roots can be weak unspecialised pathogens, entering and killing juvenile but not mature roots. The attack is usually only serious if seedling growth is slowed or stopped, resulting in damping-off of seedlings growing under unfavourable conditions. The specialised vascular wilt fungi (*Fusarium* sp. and *Verticillium* sp.) can also only attack juvenile roots, but once inside the root they escape into the vascular system.

2.4.3 Legume root nodules

Nodules are formed on the roots of actinorhizal plants by *Frankia* sp. (section 2.4.4), and on the roots of legumes by *Rhizobium* sp. and *Bradyrhizobium* sp. The bacteria obtain carbohydrate from the host plant and supply the plant with N compounds derived from atmospheric N_2. Legumes are important components of natural plant communities and agricultural systems.

Nodule formation depends upon a highly coordinated sequence of interactions between plants of the family Leguminoseae and soil bacteria belonging to the genera *Rhizobium* and *Bradyrhizobium* which results in the formation of root nodules. Only 48% of the leguminous genera have been examined for nodules to date, and of these 86% were found to be nodulated. Not all legumes are nodulated by all rhizobia. For example, lucerne (*Medicago sativa*) rhizobia only nodulate lucerne and no other legumes, and rhizobia that nodulate other legumes do not nodulate lucerne. The grouping of legumes nodulated by the same species of bacterium gave rise to the concept of cross-inoculation specificity.

In Bergey's *Manual of Systematic Bacteriology* (1984) the original classification of nodule bacteria as a single genus, *Rhizobium*, divided into species according to their cross-inoculation specificity was modified to include two genera. *Rhizobium* is now classified as a genus of fast-growing bacteria producing acid in laboratory media, which nodulate mainly temperate legumes. An example of this genus is *R. meliloti*, which forms nodules on the roots of lucerne (Figure 2.5).

The new genus, *Bradyrhizobium*, comprises slow-growing bacteria producing alkali in laboratory media, which nodulate mainly tropical legumes. An example of this genus is *B. japonicum*, which forms nodules on the roots of soybean (*Glycine max*). More recently, the genus *Azo-rhizobium* has been introduced. This at present contains only one species, *A. caulinodans*, which forms stem and root nodules on *Sesbania rostrata*.

In the non-symbiotic state rhizobia are common Gram-negative, non-

Figure 2.5 Nodule formed on the root of lucerne (*Medicago sativa*) by *Rhizobium meliloti* (magnification ×3).

spore-forming rod-shaped soil bacteria, living saprophytically on a wide range of organic C sources, but unable to fix N_2. In the presence of the appropriate legume root they are able to overcome the plant's defence mechanisms, infect the root, form nodules, synthesise nitrogenase and other vital components and fix N_2.

Despite a great deal of research into the infection of legumes by rhizobia the detailed mechanisms remain unclear. Curling and branching of root hairs are the first visible responses by the host legume, possibly due to the production of indole acetic acid by rhizobia from tryptophan excreted by the roots. A marked degree of curling by root hairs (360° or more) appears to be characteristic of a compatible interaction.

Attachment of rhizobia to the root surface may be one of the critical recognition stages prior to successful infection. Only a small and variable proportion of root hairs become infected, and many infections (68–99% in *Trifolium* sp.) abort before they reach the base of the root hair. Infection threads are tubular structures that carry the rhizobia, often in single file, from the point of entry into the root hair to the inner cells of the root cortex. Rhizobia may cause growth of the host cell wall to be re-orientated so that the root hair becomes invaginated. Alternatively, the rhizobial exopolysaccharide may stimulate an increase in plant pectic enzyme (polygalacturonase) activity which causes a softening of the root hair cell wall and allows penetration by rhizobia. Infection thread formation can be likened to a controlled incompatible host–pathogen reaction.

Not all legumes are infected via root hairs. In peanut (*Arachis hypogeae*), many mimosoid legumes and the stem nodulating *Sesbania* sp., infection occurs at emerging lateral roots, or directly through the epidermis. Because many legumes have neither root hairs, nor nodules associated with lateral roots, infection between epidermal cells may be common.

When the rhizobia penetrate the inner cortical cells of the root they multiply and cause a proportion of the cells to start proliferating thereby forming a nodule. Once the bacteria have filled a proliferating cell they become enlarged and change into bacteroids. They lose most of their ribosomes and the ability to multiply, but synthesise nitrogenase and are surrounded by membranes formed by the host cell. The bacteroids are bathed in a solution containing leghaemoglobin which transports O_2 for respiration at very low partial pressures thereby protecting the oxygen-sensitive nitrogenase. Leghaemoglobin which is similar to haemoglobin gives nodules a characteristic pink colour. The diffusion resistance of the nodule also restricts the rate of diffusion of O_2 into the nodule.

In *Rhizobium*, the genes for root hair curling, infection thread formation and host-specific nodule induction (*nod*) are located on large indigenous Sym-plasmids (see chapter 5), which also carry the genes for nitrogenase formation (*nif*) and N_2 fixation (*fix*). In *Bradyrhizobium*, the functional analogues of these *nod*, *nif* and *fix* genes are carried on the chromosome. Four of the nodulation genes, *nod* D, A, B, C are contiguous and highly conserved. These genes control root hair curling, and *nod* D is a regulatory gene which is constitutive (always produced rather than being inducible) in *Rhizobium leguminosarum* biovar *trifolii.*

The *nod* genes are regulated, via *nod* D, by compounds in the legume root exudates. In white clover, hydroxyflavones produced in the infectible zone of emerging root hairs stimulate expression of the *nod* genes within minutes, whereas coumarins and isoflavones produced behind the root tip repress *nod* gene transcription. The success of a particular infection depends upon the ratio of these stimulatory and inhibitory compounds, which provides the plant with a mechanism for regulating nodulation. Once *nod* D has interacted with the flavonoids, other *nod* genes such as *nod* F and E which are involved in host range specificity are induced, and these in turn may regulate other non-Sym plasmic genes, such as those involved in polygalacturonase production.

2.4.4 Actinorhizas

Several genera of perennial woody trees and shrubs form a N_2 fixing symbiosis with the actinomycete *Frankia.* Actinorhizal nodules occur in 21 diverse genera in eight families, and although these plants rival legumes in their contribution to global N_2 fixation, our understanding of the biology and physiology of this symbiosis is poor. Actinorhizal plants are pioneer species on sites poor in N and often exposed to extreme environmental conditions. They are often found on raw mineral soil and along streams, and in high latitude countries such as Canada and Scandinavia, actinorhizal plants thrive under conditions which are unsuitable for legumes. *Alnus* sp. is used in reclamation of material such as china clay waste and colliery spoil where trees such as *Pinus maritima* often benefit from being interplanted with *Alnus* sp.

Frankia, which was first isolated in pure culture as recently as 1978, is a genus of Gram-positive, filamentous spore-forming bacteria, which form swellings (vesicles) on the tips of the hyphae. Unlike *Rhizobium*, these organisms can fix N_2 in the absence of the host plant, and it appears that the vesicles, which form in culture and in the nodule, are the site of nitrogenase activity. Infective isolates are available for only about half of the actinorhizal genera and no species names have yet been given. *Frankia* exhibits a degree of cross-inoculation specificity.

Actinorhizal nodules, unlike legume nodules, are modified lateral roots. The hyphae penetrate deformed root hairs within a crypt formed at the junction of several root hairs. Modification of the growth of the root hair tip leads to the formation of an infection thread. Intercellular infection which does not directly involve root hairs also occurs. The endophyte proliferates within the host cortical cells which become hypertrophied with large nuclei and many mitochondria. Phenolic compounds which are inhibitory to *Frankia* accumulate in nearby uninfected cells; this may represent a partial defence response by the host plant.

Nodules may last for 3–4 years in the field and new lobes are added each year. Seasonal patterns of nitrogenase activity occur with a peak in the summer when nodules have completed their growth, and a cessation of activity in winter when the leaves of deciduous plants are lost. Spores formed in senescent nodules may form the major inoculum for soil. In the absence of the host plant the nodulation capacity of a soil decreases with time.

2.4.5 *Agrobacterium*

Agrobacterium tumefaciens is a species of soil bacterium closely related to *Rhizobium* and commonly found in the rhizosphere. It contains a large tumour-inducing (Ti) plasmid which allows the bacterium to produce neoplastic overgrowths (crown galls) on susceptible plants. Wounding of the plant is required before a sequence of processes occurs which leads to gall formation. Galls normally occur where secondary lateral roots break out from the underground stem.

The bacteria become attached to the cells of a wounded plant and anchored by cellulose fibrils produced by the bacteria. The virulence (*vir*) genes in the Ti-plasmid are induced by plant root exudates such as acetosyringone which leads to the formation of a linear single strand of T-DNA which contains the genes for synthesis of auxin and cytokinin. The T-DNA is transferred to the host plant as a protein-DNA complex; a replacement strand is produced in the bacterial donor and a complementary strand in the recipient. It is not known how the T-DNA is transported to the plant cell nucleus nor how it becomes integrated into the host genome.

Following transfer of the T-DNA, plant hormones are synthesised and rapid host cell division leads to the formation of a gall. Compounds termed opines are synthesised by the plant under the direction of the foreign DNA which are used preferentially as a substrate for growth by Agrobacterium. Opines also induce the transfer of the Ti-plasmid to other bacterial cells by conjugation.

T-DNA transfer in crown gall formation is similar to bacterial plasmid conjugation (see chapter 5); the mobilisation functions of a naturally occurring bacterial plasmid required for plasmid conjugation also allow transfer of this plasmid by *A. tumefaciens* to plants. This raises the possibility that plants may have access to the gene pool of certain soil bacteria. It is not known how widespread this ability is among bacteria and plants, however, this ability in *A. tumefaciens* makes it a useful vector for the introduction of foreign genes into plants. A non-pathogenic strain of *A. rhizogenes* has been used successfully to control crown gall formation on a wide range of crops (see chapter 5).

2.4.6 Mycorrhizas

The roots of most plant species in natural environments and in cultivation form symbiotic associations, termed mycorrhizas, with specialised fungi. This fungus–root symbiosis is, therefore, more widespread than the root nodule symbiosis which is restricted almost exclusively to certain leguminous and actinorhizal species. In nature 80% of plants have a root system which is really a mycorrhizal system. The fungus obtains carbohydrate from the plant and supplies nutrients, particularly phosphate, to the plant from the soil.

Two main types of mycorrhizas are commonly found: ectomycorrhizas and endomycorrhizas (Figure 2.6). Trees of boreal and temperate forests have ectomycorrhizas with a fungal sheath around lateral roots, intercellular penetration of the root cortex, and a mainly external vegetative mycelium. Most herbaceous and graminaceous species of temperate and semi-arid grassland, as well as many tree species in tropical and subtropical regions, have vesicular-arbuscular mycorrhizas (VA mycorrhizas). This is the most common type of endomycorrhiza (the others

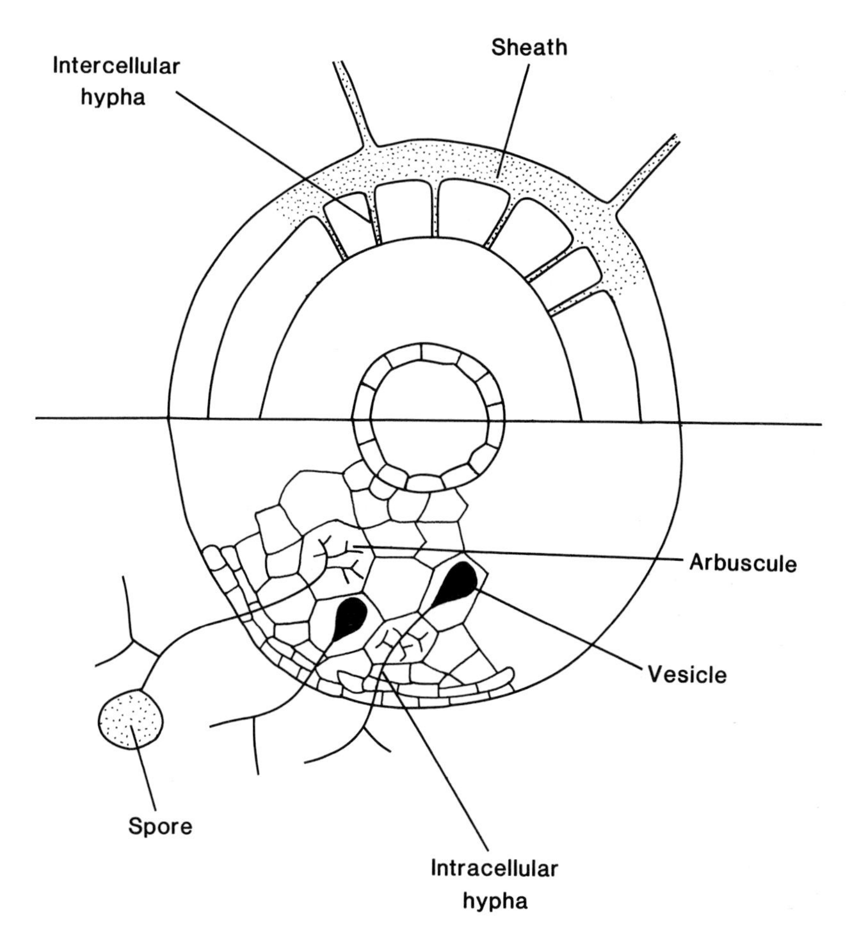

Figure 2.6 A schematic diagram of an ectomycorrhiza and a vesicular-arbuscular mycorrhiza.

are ericaceous, arbutoid and orchidaceous), with internal storage structures (vesicles), intracellular hyphal structures (arbuscules) and external branched single hyphae spreading through soil. In contrast to *Rhizobium* sp., both types of fungus show little host specificity.

Ectomycorrhizas. Ectomycorrhizas are formed mainly by Basidiomycetes, for examplee *Lactarius* sp. and *Boletus* sp. Infection may arise from existing mycorrhizal roots which act as point inoculum sources.

Mycelia fan out into the soil and when they contact an uninfected root hyphae aggregate to form strands. Mycorrhizal fungi colonising young roots are in competition with the rhizosphere microflora. Once established, a sheath forms which causes modification of root exudation. Some of the hyphae grow between the cortical cells to form the Hartig net. This provides a large surface area for the interchange of nutrients between the plant and fungus.

The fungus obtains most of its carbon from the tree in the form of sucrose which is hydrolysed extracellularly by the fungus to glucose and fructose. These are converted to fungal products (mannitol, trehalose and glycogen) which the tree cannot use, hence a one-way flow is maintained from the host plant to the fungus.

Part of this transfer of carbohydrate may represent C which would otherwise be lost from the root, for example, as exudates, but part will constitute an additional demand on the plant's photosynthate supply. A budget for Douglas fir (*Pseudotsuga douglasii*) (see chapter 3) shows that the fungal component comprises half of the annual C throughput of the stand. Data from the same stand indicate that 43% of the annual throughput of N is via the mycorrhizal fungus.

The fungal sheath (Figure 2.6) is of benefit to plants growing in soils low in nutrients because it provides a greater absorbing surface for nutrients than does the root alone. The accumulation of phosphate and other nutrients in the sheath may provide a steady supply of P etc. to the plant when transport to the root is reduced, for example during a period of drought. In addition, the relatively rapid turnover of mycelial fungus leads to a concentration of nutrients in the root region.

Mycelia are ingested by a variety of small animals such as nematodes and Collembola, and their fruiting bodies and sclerotia are important food for larger animals such as small mammals. Animals, therefore, accelerate the turnover of mycelial tissue and can disseminate mycorrhizal fungal spores.

Increased phosphatase activity of mycorrhizal roots may result in increased hydrolysis of soil organic P compounds. The fungus may also provide a degree of drought tolerance to the tree. Hyphal strands appear to link plants directly to form a continuous transport network between plants of the same and different species. This may allow the transfer of water, other nutrients, and also C, between plants.

Vesicular-arbuscular mycorrhizas. Vesicular-arbuscular (VA) mycorrhizas are formed by fungi belonging to the family Endogonaceae, for example *Glomus* sp. and *Gigaspora* sp. Roots become infected by hyphae growing through soil from propagules such as spores, or dormant fungal structures in dead previously infected root material, or from nearby roots. The large resting spores (80–150 μm diameter) are produced on the coarse external hyphae, and are borne singly in the soil or aggregated into sporocarps. Spores of *Acaulospora laevis* germinate after 1–2 months, whereas inocula consisting of infected roots can infect rapidly. Spores can germinate in the absence of plant roots but it is not clear whether the presence of roots enhances germination.

The infection process includes arrival of the fungus at the root, penetration and development of the infection, and its spread to other parts of the root. Formation of an appressorium (a swollen structure formed on the end of a spore germ tube in contact with the root) often occurs as a prelude to infection. Hyphae then penetrate the epidermal cells or pass between these cells and penetrate the outer cortical cells.

Repeated dichotomous branching of the invading hyphae within the host cell gives rise to an arbuscle (Figure 2.6). Between the plasmalemma of the host cell and the wall of the hypha is a polysaccharide matrix. Vesicles, 50–70 μm in diameter, form as inter- and intra-cellular swellings along or at the tips of hyphae. They contain lipid droplets and sometimes glycogen, and probably act as temporary storage organs. The internal structure of VA mycorrhizas can be observed in cleared and stained root samples under the light microscope. Despite the density of internal hyphal development the morphology of the root is little affected by infection. The host cell increases its cytoplasmic content and metabolic activity. In soil containing VA fungus propagules, contained development of the infection is due to spread of existing infections and to new infections.

The reasons for the lack of specificity of VA mycorrhizas are not known. Recognition might occur at the appressoria stage, but these structures are not always associated with infection. The apparent lack of infection of plants belonging to the families Chenopodiaceae, Brassicaceae and Caryophyllaceae could be due to a physical barrier at the cell wall, absence of essential nutrients, or production of toxins by the plant.

Factors such as pH, temperature, pesticides, and nutrient level affect the amount of root infected. For example, increased soil phosphate decreases VA mycorrhizal infection. Phosphate acts on the plant phase and on the soil phase (such as spore germination) of the fungi, and fungi vary in their sensitivity to increased phosphate. Phosphate deficiency causes increased exudation from roots and this may stimulate mycorrhizal infection. To date it has not been possible to grow VA mycorrhizal fungi in pure culture. Growth of the fungus in the symbiotic association may be due to the host plant providing some unknown growth factors in addition to energy sources.

VA mycorrhizal plants generally grow better in phosphate-deficient soils than non-mycorrhizal plants. Studies using ^{32}P tracers have shown that mycorrhizal plants use the same pool of soil phosphate as non-mycorrhizal plants. The fungus therefore extends the volume of soil from which the plant can take up phosphate rather than solubilising otherwise unavailable forms of P. This is important for an immobile nutrient such as phosphate which is depleted in the soil around roots, but less important for a mobile nutrient such as NO_3^-. Inorganic P is concentrated by a factor of a thousand in the mycorrhizal hyphae and is stored and transported in the form of polyphosphate. It is not clear how phosphate is transferred into the host plant from the fungus, nor how carbohydrate is transferred to the fungus from the plant.

Mycorrhizas may also increase the uptake of other nutrients such as S, Cu, Zn and NH_4^+ and allow plants to grow in harsh conditions such as heavily polluted, acidic and eroded sites. Mycorrhizas sometimes increase resistance to disease and drought and can increase N_2 fixation by legumes. Many of the beneficial effects of mycorrhizas, for example improved growth in dry soils, may be indirect due an improvement in the mineral nutrition of the plant.

2.4.7 Root-knot nematodes

Root-knot nematodes (*Meloidogyne* sp.) are widely distributed in soils, have an extensive host range and are estimated to account for an annual loss in crop yield of 5% worldwide. Losses due to this pest are particularly large on small farms in developing countries. There are four main pest species: *M. incognita*, *M. javanica*, *M. arenaria*, and *M. hapla*. They survive in soil as eggs or larvae, and are disseminated by water, animals and plants.

Larvae develop inside the eggs, which hatch, producing second-stage juveniles which migrate through soil and are attracted to roots. On contact, the juveniles penetrate the root, using stylets and enzymes, just below the root tip, and move through the root to near the vascular tissue. This causes an increase in plant cell size and galls and knots (1–10 mm in diameter) form on the roots. The juvenile forms change into mature adults, completing their life cycle in 17–57 days. Intensive knot formation causes root deformation, decreased efficiency of uptake of water and nutrients and premature defoliation and death of plants.

2.5 Summary

This then completes our review of the major players in the world of the soil. In order to understand the roles of these organisms in soil it will be necessary in the next chapter to consider some basic facts of life.

3 Biological processes in soil

3.1 Introduction

This chapter examines the role of the soil population in the transformation of materials in soil, in particular the elements C, N, P and S. The coverage is restricted; there are many excellent texts giving details of nutrient cycling (see references and further reading). What it attempts to do, however, is to show how transformations of elements in soil can be viewed from the perspective of the nutrition of the soil population, their requirements and how these are satisfied. The subsequent chapters consider the implications of these processes in terms of soil development and environmental issues.

3.2 What do organisms require for life?

A useful starting point when discussing biological processes in soil is to consider the main nutrients needed by soil organisms. In other words, if a bacterium is to live in soil what are its basic requirements for life? These are summarised in Figure 3.1.

Of the total cell weight, 80–90% is water; it is normal practice to express the elemental content of materials such as bacteria and plant shoots on a dry-weight basis after removing most of the water in an oven at about 80°C. Examples of macronutrients are C, N, P and S. Micronutrients include Fe, Cu and Mo, and some organisms require pre-formed compounds (growth factors) such as vitamins. Energy, C and N are now considered in more detail.

3.2.1 Energy and carbon

Organisms employ three methods for obtaining energy: fermentation, respiration, and photosynthesis. In all these processes the harvested energy is stored in the form of adenosine triphosphate (ATP) and used for biosynthesis. ATP is generated as electrons are transported by reduction–oxidation reactions via a series of compounds to a terminal electron acceptor (Figure 3.2).

The flow of electrons across the cytoplasmic membrane is coupled to a net translocation of protons from the cytoplasm to the external aqueous

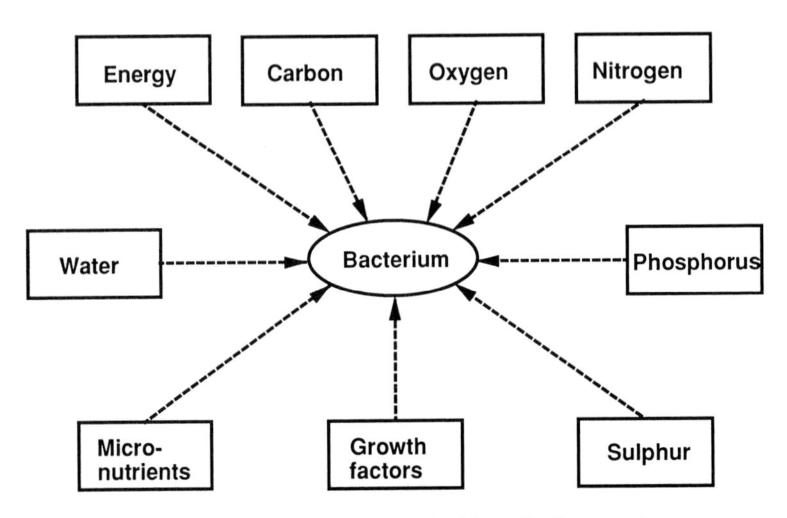

Figure 3.1 The main nutrients required by soil microorganisms.

phase. The resulting proton electrochemical gradient (also known as the proton motive force) drives protons through the ATP synthase enzyme associated with the membrane, thereby generating ATP. The proton electrochemical gradient also drives active transport to and from the cell, the movement of flagella being used to propel bacteria and reverse (endergonic) electron transport (section 3.3.2).

Fermentation (substrate level phosphorylation) is the simplest method of ATP generation in which organic compounds serve as electron donors (are oxidised) and electron acceptors (are reduced). However, the average oxidation state of the products is the same as that of the substrates. It can proceed in the absence of O_2, but the process is relatively inefficient and yields of ATP are only 2 mol per mol of glucose. Carbohydrates are the main substrates yielding products such as acetic acid, but bacteria can also ferment organic acids and amino acids.

Respiration (oxidative phosphorylation) is a more complex and efficient method of generating ATP in which organic or inorganic compounds serve as electron donors (are oxidised) and inorganic compounds serve as the terminal electron acceptors (are reduced) as shown in Figure 3.2. The use of NH_4^+ and NO_2^- as electron donors forms the basis of the process of nitrification.

Under aerobic conditions O_2 serves as the terminal electron acceptor (aerobic respiration), but certain bacteria can carry out respiration in the absence of O_2; under anaerobic conditions NO_3^-, SO_4^{2-} or CO_2, and other compounds may serve in place of O_2 (anaerobic respiration). The use of

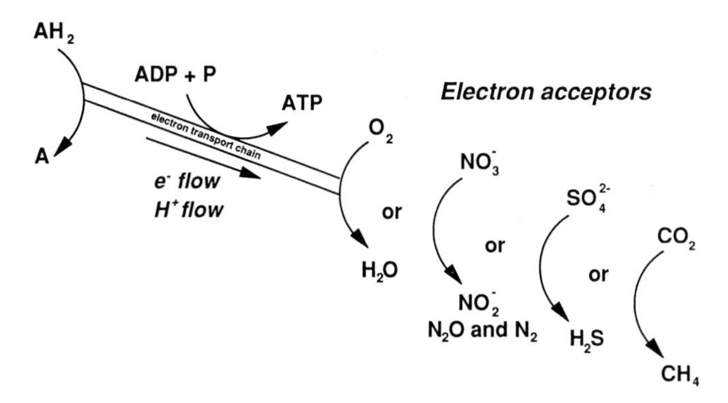

Figure 3.2 Schematic diagram of transfer of electrons from a donor to alternative acceptors during aerobic respiration or anaerobic respiration. After J.F. Wilkinson (1986) *Introduction to Microbiology*, Blackwell Scientific Publications, Oxford.

NO_3^- or SO_4^{2-} as terminal electron acceptors forms the biochemical basis of the important soil processes of denitrification and sulphate reduction, respectively (sections 3.3.3 and 3.3.5). The use of CO_2 under highly reducing conditions leads to methane (CH_4) production. Some organisms can grow either in the absence or presence of O_2 (facultative anaerobes) and some can only grow in the absence of O_2 (obligate anaerobes). Aerobic respiration yields 38 mol ATP per mol of glucose.

In photosynthesis (photophosphorylation), light energy is absorbed by a photosynthetic pigment system, including chlorophyll, and electrons (and protons) are transported from H_2O, the electron donor, which leads to O_2 production. Organisms which obtain energy by this method are termed phototrophs, to distinguish them from those organisms which obtain energy from chemical sources by respiration and fermentation (chemotrophs).

Organisms may obtain C from either organic compounds (heterotrophs) or from CO_2 (autotrophs). Fixation of CO_2 by microorganisms involves the same enzymes, for example ribulose bisphosphate (RuBP) carboxylase, as in plants. Organisms may therefore be grouped according to the sources from which they obtain their energy and C (Figure 3.3). Photoautotrophs, for example maize (*Zea mays*), obtain energy from light and C from CO_2 and photoheterotrophs, for example the green bacterium (*Chlorobium* sp.) use light energy and organic C. Chemoautotrophs (also termed chemolithotrophs) obtain energy from chemical reactions and C from CO_2, for example the bacterium *Nitrosomonas europea* obtains energy from the

Energy source	Carbon source CO_2 *autotroph*	Carbon source Organic C *heterotroph*
Light *Photo-*	*Photoautotroph* **Algae** **Plants**	*Photoheterotroph* **Green and purple bacteria**
Chemical *Chemo-*	*Chemoautotroph* **Nitrifying and S oxidising bacteria**	*Chemoheterotroph* **Most bacteria** **Fungi** **Protozoa**

Figure 3.3 Groupings of soil organisms according to their source of C and energy.

oxidation of NH_4^+ to NO_2^- and *Nitrobacter* sp. from the oxidation of NO_2^- to NO_3^-; these reactions constitute nitrification, an important component of the N cycle in soils (section 3.3.2). Chemoheterotrophs, for example the fungus *Trichoderma harzianum* obtain both energy and C from organic compounds.

3.2.2 Nitrogen

The preferred source of N for bacteria is NH_4^+, although organisms containing nitrate reductase (a molybednum-containing enzyme which reduces NO_3^- to NH_4^+) can also assimilate NO_3^-. Assimilation of NO_3^- is a highly regulated process which normally proceeds slowly at the rate at which NH_4^+ is required for growth. The intermediate NO_2^- rarely accumulates; this contrasts with dissimilatory nitrate reduction (section 3.3.2) which is more rapid, and which can lead to an accumulation of NO_2^-. Plants can take up either NH_4^+ or NO_3^-, but where the supply of NO_3^- is limited, for example in acid soils, then organisms may utilise mainly NH_4^+. Heterotrophic organisms may obtain N from organic compounds such as peptides and amino acids. A single compound such as an amino acid can provide a microorganism with energy, C and N. A few genera of bacteria are able to utilise atmospheric N_2 gas (section 3.3.1).

All sources of N are converted into NH_4^+, and then into glutamate and glutamine, which are key intermediates in subsequent biosynthesis of N compounds. It is not clear how ions such as NO_3^- and SO_4^{2-} are taken into a cell which has a membrane potential which is negative inside the cell.

3.3 Selected biochemical processes

All biochemical processes rely upon catalysts to accelerate the rate of reaction. Biological catalysts are termed enzymes. They are proteins and, like inorganic catalysts, remain unchanged after completion of the reaction (soil enzymes are discussed in more detail in section 3.4). The sequence of amino acids which make up such a protein is determined by the genetic information carried in the cell in the form of deoxyribonucleic acid (DNA). A unique sequence of bases in the DNA (a gene) codes for a particular amino acid. Some genes influence the expression of other genes, thereby controlling enzyme synthesis or activity. Some of the biochemical processes which have been mentioned in relation to the nutrition of soil bacteria are now considered in more detail. In some cases details of the genetics of the processes are included.

Although it is possible to present nutrient transformations as simple cycles (e.g. the N cycle and S cycle), detailed investigation of the biochemical pathways for the processes and the ecology of the organisms have revealed a more complex nature to the transformation of elements in soil. Processes have been selected here which are particularly important in the transformation of N and S in soil.

3.3.1 Nitrogen fixation

Biological nitrogen fixation is a process of enormous global importance, yet the ability to exploit the vast pool of dinitrogen gas in the atmosphere is restricted to only a few genera of prokaryotic organisms. These include, among others, the free-living aerobic bacterium *Azotobacter* sp., the obligate anaerobe *Clostridium* sp., the photosynthetic blue green bacterium *Anabaena* sp., and the symbiotic bacteria *Rhizobium* sp. and *Frankia* sp.

The process, which provides the organism with a supply of nitrogen, can be represented simply by the following equation:

$$N_2 + 6H^+ + 6e^- \rightarrow 2NH_3$$

The N_2 molecule is extremely stable; the industrial production of NH_3 for fertiliser from N_2 by the Haber–Bosch process requires a temperature of 400°C and a pressure of 20–35 MPa (200–350 atmospheres), together with a finely divided Fe catalyst, to break the triple covalent bond between the N atoms. Certain bacteria are able to carry out this reaction at normal temperature and pressure using an enzyme. This enzyme, termed nitrogenase, is similar in all N_2-fixing organisms. It consists of two proteins, a large one containing Fe and Mo, and a small one containing Fe, both of which are destroyed by O_2.

Nitrogen-fixing organisms growing in an aerobic environment have evolved a variety of mechanisms to protect nitrogenase from O_2, for example the production of leghaemoglobin in legume root nodules (see chapter 2). The free-living bacterium *Azotobacter* sp. maintains a low intracellular O_2 concentration by having a high rate of respiration which can be uncoupled from ATP generation to prevent excess ATP production. In photosynthetic N_2-fixing organisms this problem is exacerbated by the production of O_2 during photosynthesis. Cyanobacteria (blue-green bacteria) have specialised cells (heterocysts) in which N_2 fixation occurs but which do not photosynthesise. Fixed C and N are transported to and from heterocyst cells.

The biological reduction of N_2 also requires a large amount of energy; 12–15 ATP molecules are required for the reduction of one molecule of N_2. ATP in the form of MgATP binds to the Fe-protein and stimulates transfer of electrons obtained from reduced ferredoxin and flavodoxin to Fe atoms in the MoFe-protein. At least one ATP is hydrolysed for each electron transferred. Nitrogen binds to the active site of the MoFe-protein and is reduced to a series of enzyme-bound dinitrogen hydride intermediates (the exact identity of these is unknown) and eventually to ammonia, which is assimilated as glutamate and glutamine.

It has been suggested that N_2 fixation imposes a high cost on a legume in terms of the energy (and therefore the amount of photosynthate) needed to assimilate N compared to the assimilation of inorganic N from soil or fertiliser. From biochemical studies estimates have been obtained of the theoretical energy cost (expressed in terms of amounts of organic C required) for these two forms of N assimilation by plants.

N_2 fixation 2.9–6.1 g C per g N
Nitrate use 0.8–2.4 g C per g N

These data indicate a slight energetic advantage in favour of nitrate use. However, measurements of amounts of C used for N assimilation by fixing and non-fixing plants show no clear differences. Also, field experiments in India and Australia using soyabean plants either fixing N_2 or using nitrate have shown no differences in dry matter production, which is an indicator of net energy yield. Therefore, N_2 fixation imposes no burden, compared to fertiliser N, in terms of crop yield. Note that this comparison takes no account of the origin of the energy used in providing these two different sources of N; this is discussed further in chapter 6.

Nitrogenase also reduces other triple-bonded molecules such as acetylene, in this case producing ethylene. This forms the basis of the acetylene reduction assay which has been used extensively to measure rates of N_2 fixation. Problems associated with the use of this method to measure rates of N_2 fixation in the field include endogenous ethylene production by soil microorganisms, inhibition of hydrogen evolution from nodules and non-linear production of ethylene with time. Such measurements provide only short-term estimates of nitrogenase activity and large errors are associated with extrapolation of data to seasonal estimates.

The only direct measurement of N_2 fixation is given by $^{15}N_2$ reduction in which a legume or other N_2-fixing system is incubated in an atmosphere enriched with the stable isotope ^{15}N. The amount of isotope fixed is measured by mass spectrometry. This method is expensive and not suitable for use in the field, but it is essential for calibration of other techniques. A variety of indirect methods has been developed based upon the principle of ^{15}N isotope dilution for estimating N_2 fixation and other N transformations in soil.

Nitrogenase can also reduce H^+ which leads to H_2 evolution; at least one molecule of H_2 is produced for each molecule of N_2 reduced. The reaction catalysed by nitrogenase should therefore be represented as follows:

$$N_2 + 8H^+ + 8e^- \rightarrow 2NH_3 + H_2$$

It is also known that some forms of nitrogenase exist which contain vanadium rather than molybdenum, although the detailed chemistry of this enzyme is not known.

The genetics of N_2 fixation have been intensively studied. The production and regulation of nitrogenase is controlled by a closely linked series of 17 *nif* genes. *nif* D and *nif* K control synthesis of the MoFe proteins and *nif* B is involved in the construction of FeMo cofactors. *nif* A is a regulatory gene which may be activated by plant compounds such as flavonoids, similar to the *nod* D gene (see chapter 2).

3.3.2 Nitrification

Nitrification is a process used by only a few genera of autotrophic bacteria as a means to generate energy. The energy yield for the reaction is low, resulting in very slow growth rates. The overall process, which involves two reactions, can be represented by the following overall equation:

$$NH_4^+ \rightarrow NO_2^- \rightarrow NO_3^-$$

Ammonium oxidation is restricted to five genera of bacteria, for example, *Nitrosomonas* sp. The energy yield is only 272 kJ per mole of NH_4^+ oxidised compared to the oxidation of glucose which yields 2872 kJ per mole. Hydroxylamine (NH_2OH) is an intermediate in the reaction. The overall reaction generates acidity as excess protons are released:

$$NH_4^+ + O_2 \rightarrow NO_2^- + 4H^+ + 2e^-$$

NH_4^+ contains N in its most reduced state, but it is not readily oxidised. The direct oxidation of NH_4^+ to NH_2OH by O_2 is endergonic, and therefore requires a special type of enzyme or chemically reactive species. Once the first N–O bond is formed the subsequent oxidation is more favourable which allows NH_4^+ oxidation to NO_2^- to be used as an energy source.

In autotrophic nitrifying bacteria hydroxylamine is formed by ammonium monooxygenase (a copper-containing enzyme):

$$NH_4^+ + O_2 + H^+ + 2e^- \rightarrow NH_2OH + H_2O$$

The reaction is a reduction and relies upon electrons produced by the subsequent oxidation of NH_2OH:

$$NH_2OH + H_2O \rightarrow NO_2^- + 5H^+ + 4e^-$$

A small proportion of the electrons transported across the membrane by this reaction are transported back (reverse electron transport) to generate reductants to allow the conversion of NH_4^+ to NH_2OH. This is an unusual reaction because the cell relies upon the monooxygenase to produce hydroxylamine which is the electron donor for the reaction.

Nitrite oxidation, which is even less energetically favourable (71 kJ per mole of NO_2^-), is restricted to one genus of bacteria, Nitrobacter:

$$NO_2^- + H_2O \rightarrow NO_3^- + 2e^- + 2H^+$$

Reverse electron transport of approximately 2% of the electrons produced from NO_2^- is used to generate reductant (NADH) which may be used, for example, in assimilation of C from CO_2.

There is evidence that some heterotrophic organisms, both bacteria and fungi, also oxidise NH_4^+ and certain organic N compounds to produce NO_2^- and NO_3^- although the biochemical basis of this reaction is not known. It may be that the free radicals produced during the enzymatic degradation of lignin (section 3.4.4) also oxidise NH_4^+. In the laboratory, methanotrophs (bacteria able to grow on methane) are able to oxidise NH_4^+ using methane monooxygenase, but the extent to which this occurs in the field is not clear. Heterotrophic nitrifiers such as the fungus *Absidia* sp. may be important in acid forest soils.

3.3.3 Denitrification

Denitrification involves a series of reactions used by a wide range of facultative anaerobic bacteria in the absence of O_2, in which oxidised forms of N are used as alternative electron acceptors. The overall reaction involves dissimilatory reduction of N oxides to nitrogen gases:

$$NO_3^- \rightarrow NO_2^- \rightarrow (NO) \rightarrow N_2O \rightarrow N_2$$

Specific reductase enzymes have been isolated for each of the following stages, except for nitric oxide (NO) reduction:

$$NO_3^- + 2e^- + 2H^+ \rightarrow NO_2^- + H_2O$$
$$NO_2^- + 2e^- + 2H^+ \rightarrow NO + H_2O$$
$$2NO + 2e^- + 2H^+ \rightarrow N_2O + H_2O$$
$$N_2O + 2e^- + 2H^+ \rightarrow N_2 + H_2O$$

The enzyme responsible for the first stage of denitrification, dissimilatory nitrate reductase, is an Fe–S protein containing Mo (similar to nitrogenase). In *Pseudomonas aeruginosa* and *Rhizobium* sp. distinct assimilatory and dissimilatory nitrate reductase enzymes have been identified. Dissimilatory nitrate reductase is repressed in the presence of O_2 and derepressed in the absence of O_2. This mechanism allows O_2 (which gives a higher molar energy yield than NO_3^-) to be used when available, also sparing NO_3^- for assimilation. However, the concentration at which derepression occurs varies with different organisms. In soil, where the concentration of O_2 may be continuously changing, it may be an advantage for an organism to retain dissimilatory nitrate reductase activity under aerobic conditions to allow a rapid response to anaerobiosis. Under certain conditions, for example at low pH values, the reaction may stop at nitrous oxide (N_2O).

The bacteria capable of denitrification comprise some 13 genera including *Pseudomonas* sp., *Agrobacterium* sp. and *Bacillus* sp.. *Bradyrhizobium japonicum* is capable of denitrification but it is not clear whether this process is linked in any way to N_2 fixation. *Thiobacillus denitrificans* uses NO_3^- as the terminal electron acceptor during oxidation of reduced sulphur compounds such as sulphide.

Under conditions of reduced O_2 concentration (but not necessarily completely anaerobic) in the presence of the appropriate bacteria and a supply of readily available organic substrate, significant quantities of N may be lost from the soil by denitrification. This is of concern as N_2O is one of the so-called greenhouse gases (see chapter 5).

N_2O may also be produced during nitrification accounting for approximately 0.1% of the N oxidised. The enrichment of ^{18}O and ^{15}N in N_2O produced from nitrification is lower than the enrichment of N_2O in the atmosphere, which suggests that the production of this gas during nitrification, probably via NO_2^-, is a minor source of atmospheric N_2O compared to denitrification. Nitrite reduction may provide ammonium oxidisers with a strategy for dissipation of nitrite in circumstances where it may otherwise accumulate to toxic concentrations.

3.3.4 Sulphur oxidation

Sulphur oxidation is a reaction used by one genus of autotrophic bacteria, *Thiobacillus*, as a means to obtain energy. It can be represented by the following overall reaction:

$$S^{2-} \rightarrow SO_3^{2-} \rightarrow SO_4^{2-}$$

Five species of *Thiobacillus* have been studied in detail and each one can oxidise a range of reduced S compounds including elemental S. All are aerobic except for *T. denitrificans*; *T. thiooxidans* and *T. ferrooxidans* have optimum pH values of 2–3.5 for growth. The oxidation of elemental sulphur by *T. thiooxidans* leads to the production of acidity:

$$2S + 3O_2 + 2H_2O \rightarrow 2H_2SO_4$$

Thiobacillus ferrooxidans can oxidise either reduced S compounds or ferrous (Fe^{2+}) ions, also generating acidity:

$$2FeS_2 + 7O_2 + 2H_2O \rightarrow 2FeSO_4 + 2H_2SO_4$$

The oxidation of S compounds involves adenosine-5′-phosphosulphate (APS) and 3′-phospho-adenosine-5-phosphosulphate (PAPS):

$$\begin{aligned} 2AMP + 2SO_3^{2-} &\rightarrow 2APS + 4e^- \\ 2APS + PO_4^{3-} &\rightarrow 2ADP + 2SO_4^{2-} \end{aligned}$$

Heterotrophic organisms, both bacteria and fungi, can also oxidise S compounds, but they do not obtain energy from the reaction.

3.3.5 Reduction of inorganic sulphur

SO_4^{2-} is reduced during the assimilation of SO_4^{2-} by bacteria. The assimilatory reductive pathway is closely regulated to meet the requirement for the biosynthesis of cysteine. A few genera of bacteria are also able to use SO_4^{2-} and other oxidised forms of S as alternative electron acceptors for anaerobic oxidation of H_2 and organic compounds (dissimilatory SO_4^{2-} reduction). No significant regulation of the enzymes involved in dissimilatory SO_4^{2-} reduction has been observed.

The obligate anaerobic bacterium *Desulfovibrio desulfuricans* has received most attention, although *Clostridium* sp. and *Desulfotomaculum* sp. are also important. The overall process can be represented by the following equation:

$$SO_4^{2-} \rightarrow SO_3^{2-} \rightarrow S_3O_6^{2-} \rightarrow S_2O_3^{2-} \rightarrow S_2^-$$

The reduction of SO_4^{2-} to SO_3^{2-} involves ATP sulphurylase which catalyses the following reaction:

$$SO_4^{2-} + \text{ATP} \rightarrow \text{APS} + \text{PP}_i$$

and APS reductase:

$$\text{APS} + 2e^- \rightarrow \text{AMP} + SO_3^{2-}$$

The SO_3^{2-} is then protonated by a chemical reaction to form bisulphite (HSO_3^-), and subsequent reduction reactions are catalysed by bisulphite reductase, trithionate ($S_3O_6^{2-}$) reductase and thiosulphate ($S_2O_3^{2-}$) reductase to produce sulphide (S_2^-).

Hydrogen sulphide (H_2S) produced as a result of this process under anaerobic conditions is toxic to most aerobic organisms including crop plants, and may cause precipitation of metal sulphides and hence the characteristic black colour of anaerobic muds.

3.4 Soil enzymes

The biochemical versatility of the soil bacterial population provides soil with the capability to degrade all natural compounds and most of the synthetic compounds which enter either deliberately, for example as pesticides to crops, or accidentally, for example industrial pollution of land. Many of the enzymes involved in these reactions are located within the organisms (endocellular enzymes). However, soils also possess enzyme activity which persists after the microbial population has been inhibited or killed. These are termed extracellular or abiotic enzymes.

Extracellular enzymes are mainly derived from soil microorganisms, particularly those enzymes involved in the degradation of insoluble substrates such as proteins and carbohydrates which are too large to enter the cell and which must therefore be at least partially broken down outside the cell. Plants and animals may also produce extracellular enzymes, but there are problems in clearly distinguishing between enzymes produced by these organisms themselves and enzymes produced by associated micro-organisms in the rhizosphere and in the animal gut.

The accumulation of extracellular enzymes in soil depends upon mechanisms for the stabilisation of these molecules, which being proteins are themselves liable to degradation by proteases. Enzymes may be adsorbed to the surface of clay minerals and between clay lamellae. Enzymes associated with organic matter are more resistant to degradation, possibly due to changes in viscosity or H bonding. The association of an enzyme with a solid surface such as a clay particle may lead to changes in substrate affinity due to unfolding or coagulation of the protein chains, or physical blocking of active sites. Alternatively, enzymes may become aligned in a way that enhances activity. A hypothetical scheme has been proposed for the stabilisation of enzymes within soil (Figure 3.4). A highly protected enzyme will be of no significance to soil enzyme activity if diffusion of substrates to the enzyme and products away from the enzyme are not possible.

Results of studies on extracellular enzyme activity in soils are highly dependent upon the methods used. In order to study such enzymes, soil endocellular activity must be inhibited without affecting the chemical and physical conditions and the extracellular enzyme activity. Antibiotics, high energy ionising radiation and a range of antiseptic and bacteriostatic agents have been used; one of the most acceptable is toluene. Furthermore, soil enzyme activity is often measured under standard but artificial conditions, for example as buffered suspensions at pH 7 with either the natural substrate or a synthetic substrate provided in excess. Such assays give an estimate of the maximum potential enzyme activity, but not the activity which actually occurs in soil.

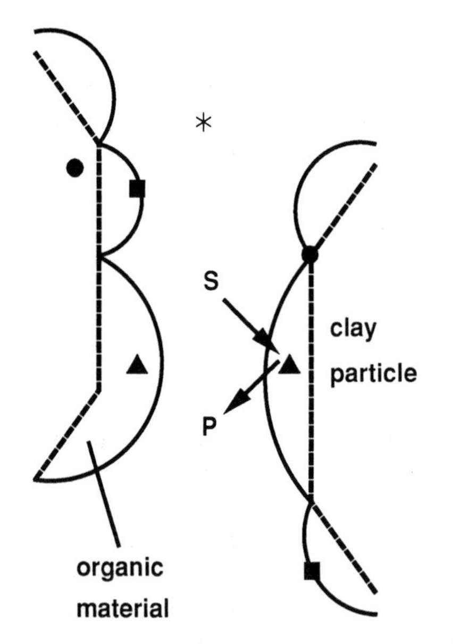

ENZYME LOCATIONS

▲ trapped in a film of organic material

■ attached to surface of organic material film

● associated with clay (surface or interlamellar)

* free enzyme in solution

S substrate

P product

Figure 3.4 A schematic diagram for the location of extracellular enzymes in soil. After R.G. Burns (1977).

A few of the major soil enzymes which exhibit significant extracellular activity are now considered.

3.4.1 Carbohydrases

The activities of carbohydrases such as cellulase are relatively easy to measure because the substrate is often insoluble, the reaction products are soluble, and neither reacts with soil particles. In common with many soil enzymes the activity of invertase (which catalyses the hydrolysis of sucrose to glucose and fructose) decreases down the soil profile, but varies considerably, and there is no correlation between activity and other soil properties.

There are many cellulolytic fungi, for example *Aspergillus* sp., but fewer cellulolytic bacteria, although actinomycetes such as *Streptomyces* sp. produce cellulase. Cellulase is not a single enzyme but is an enzyme system, and it is normally inducible, with simple glucose-containing molecules such as cellobiose acting as inducers.

3.4.2 Esterases

Phosphatase is important, particularly in soils with low levels of available P, in the mineralisation of inorganic P from organic sources such as inositol phosphate. The measurement of phosphatase activity is complicated because inorganic P reacts with soil components and may not therefore be detected. The release of phenol from phenylphosphate is a more reliable assay for this enzyme. Different phosphatases have different optimum pH values; there are acid, neutral and alkaline phosphatases. When organisms become short of P they may increase the production of phosphatase and hence the supply of inorganic P.

Nucleases, which degrade nucleic acids, are also found in soils. However, if nucleic acids become complexed with lignin they are more resistant to enzyme attack.

3.4.3 Proteases and amidases

Proteases and amidases accumulate in soil and are important in the degradation of proteins and therefore in the mineralisation of N. Urease is the soil enzyme which has been studied in most detail, mainly due to the widespread use of urea as a N fertiliser. Urease is stable and persistent in soil and extracellular activity may account for 50–90% of the total soil urease activity. Table 3.1 shows the urease activity measured in soils from a range of countries. The variation in activity measured for a single soil is greater than the variation between soils.

Table 3.1 Urease activity (μg urea hydrolysed per gram of soil per hour) in a range of soils. After J.M. Bremner and R.L. Mulvaney (1978), in *Soil Enzymes*, ed. R.G. Burns, Academic Press, pp. 149–196

Location	Organic C (%)	Soil pH	Urease activity
Australia	0.16–5.88	4.8–6.7	22–416
USA	0.30–6.73	4.6–8.0	11–189
Malaysia	0.64–3.44	4.2–4.9	5–41
India	0.24–0.98	4.2–6.7	20–162

When urea fertilisers are applied to soil, ammonia is produced by enzymatic hydrolysis:

$$CO(NH_2)_2 + H_2O \rightarrow CO_2 + 2NH_3$$

Under alkaline conditions NH_3 may be lost from the soil by volatilisation. Similarly, NH_3 may volatilise from localised areas of urine deposited on the soil surface by grazing animals. It has been estimated that on average sheep void 45 g of urea per animal per day and cattle void 140 g of urea per animal per day. The addition of urease inhibitors such as hydroquinone to soil may reduce the losses of N by this process.

3.4.4 *Oxidoreductases*

Lignin is a complex polymer of aromatic nuclei, containing a basic phenyl propane unit grouped into a highly branched random structure linked by strong bonds. It is therefore extremely resistant to chemical and microbial degradation, although certain fungi, in particular white rot fungi, are ligninolytic. The enzyme responsible for lignin degradation (ligninase) was isolated in 1983 from the white rot fungus *Phanerochaete chrysosporium*. It contains a haem group which reacts with hydrogen peroxide to form a molecule with a high affinity for electrons which then oxidises the aromatic rings in lignin to form relatively stable free radical cations. These radicals can bring about further oxidation of the lignin molecule at sites remote from the enzyme, and eventually cause the degradation of lignin.

Dehydrogenase is an enzyme common to most microorganisms, but in contrast to the other enzymes discussed so far it is predominantly endocellular. The amount of dehydrogenase activity in soil has been used as an indicator of the activity of the soil microbial population. 2,3,5-triphenyl-tetrazolium chloride (TTC), a pale-coloured water-soluble compound, acts as an electron acceptor for many dehydrogenases and is reduced to triphenyl formazan, a red-coloured water-insoluble compound soluble in methanol, which can be measured colorimetrically.

3.5 Sources of substrates for heterotrophs

The importance of photosynthesis in providing a supply of energy-rich organic materials to the soil has been considered in chapter 2. Plant material enters the soil by a variety of routes (Figure 3.5). In natural ecosystems shoot material (litter) forms the major input of organic material. If it is assumed that an ecosystem is at steady state (i.e. the amount of plant biomass is neither increasing or decreasing) then net primary production (estimated from shoot data) gives the potential input of organic material into the soils. Some of this enters the soil indirectly, following grazing by animals, as excreta or corpses. For example, in grazed upland areas of the UK up to 80% of the organic N fixed by white clover (*Trifolium repens*) and returned to the soil passes through the grazing sheep. The relatively resistant soil organic matter will also provide substrates, although the distinction between resistant plant and animal material and humic material is not clear.

Where annual crops such as wheat (*Triticum aestivum*) or beans (*Phaseolus vulgaris*) are grown, most of the shoot material is removed from the land at harvest, particularly in tropical regions, therefore the input of organic material is largely restricted to that derived from roots. This includes material lost from the roots while the plants are growing and from the dead roots remaining at harvest. The practical difficulties of studying

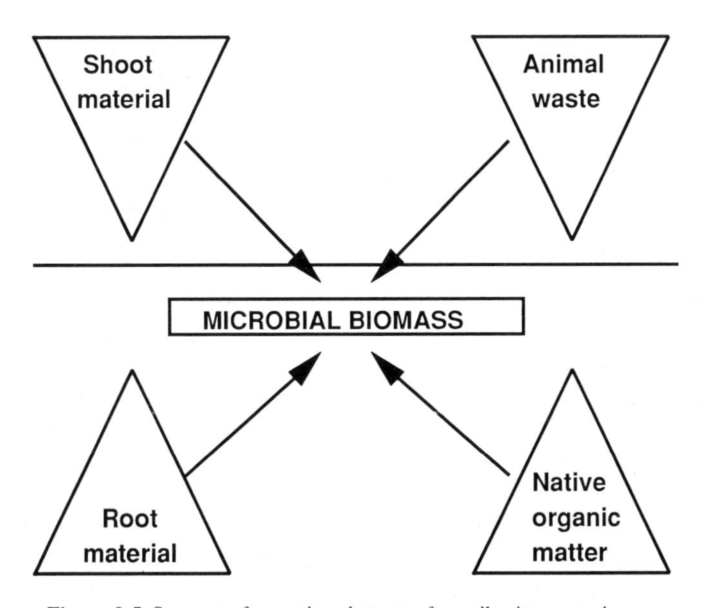

Figure 3.5 Sources of organic substrates for soil microorganisms.

roots growing in soil means that their contribution to the input of organic material into soils in the field is uncertain.

The amount of C lost from roots has been measured under laboratory conditions for cereals growing in solution and sand culture, and in soil using ^{14}C tracer techniques. Estimates for total C lost (organic C plus CO_2) from plants supplied with ^{14}C-labelled CO_2 range from 14% to 40% of total net C fixed by photosynthesis (i.e. allowing for shoot respiration). Some of the C loss measured by this technique may be in whole roots not removed from soil, which cannot be considered as rhizosphere. Loss of material is generally greater in non-sterile than sterile systems and for roots growing in solid media rather than in solution culture. Losses are also increased by anaerobiosis, moisture stress, low temperatures, removal of shoots, and herbicides. Most work has been on seedling plants, however, the nature and amount of material may change as the plant ages.

Organic materials released by roots into soil have been classified according to their origin in the root (Figure 3.6), although they are not always well-defined compounds from a single origin. Exudates are low molecular weight compounds which leak from all cells either into intercellular spaces and then into the soil through cell junctions, or directly through epidermal cell walls into the soil. Secretions comprise both low molecular weight compounds and high molecular weight mucilages which are released actively by the plant from the root cap and from epidermal cells. There are mucilages produced by bacterial degradation of the outer primary cell wall of dead epidermal cells.

Mucigel is a layer of gelatinous material on the root surface which is visible under the transmission electron microscope (Figure 2.3c); it comprises natural and modified plant mucilages, bacterial cells and their metabolic products such as capsules and slimes, and also colloidal soil mineral and organic matter. Finally, lysates are compounds released from the autolysis of older epidermal cells following failure of the plasmalemma. These cells are then colonised by microorganisms causing further release of materials into the rhizosphere.

There is little information on losses of root material from trees. The formation of an ectomycorrhizal sheath around the roots of some trees such as beech (*Fagus sylvatica*) may represent an adaptation by the fungus to maximise its utilisation of tree root exudates. If this is so then the mycorrhizal fungal biomass associated with trees is an indication of the loss of organic material from tree roots. Estimates of fungal biomass C production as great as 1541 g m^{-2} per year have been reported (Table 3.2).

Field measurements of losses of organic material from roots are more difficult to make particularly when C fluxes must be partitioned into those due to root respiration, uptake of CO_2 by autotrophic organisms (a minor component) and respiration by heterotrophic microorganisms utilising either root-derived material or existing soil organic matter (Figure 3.7).

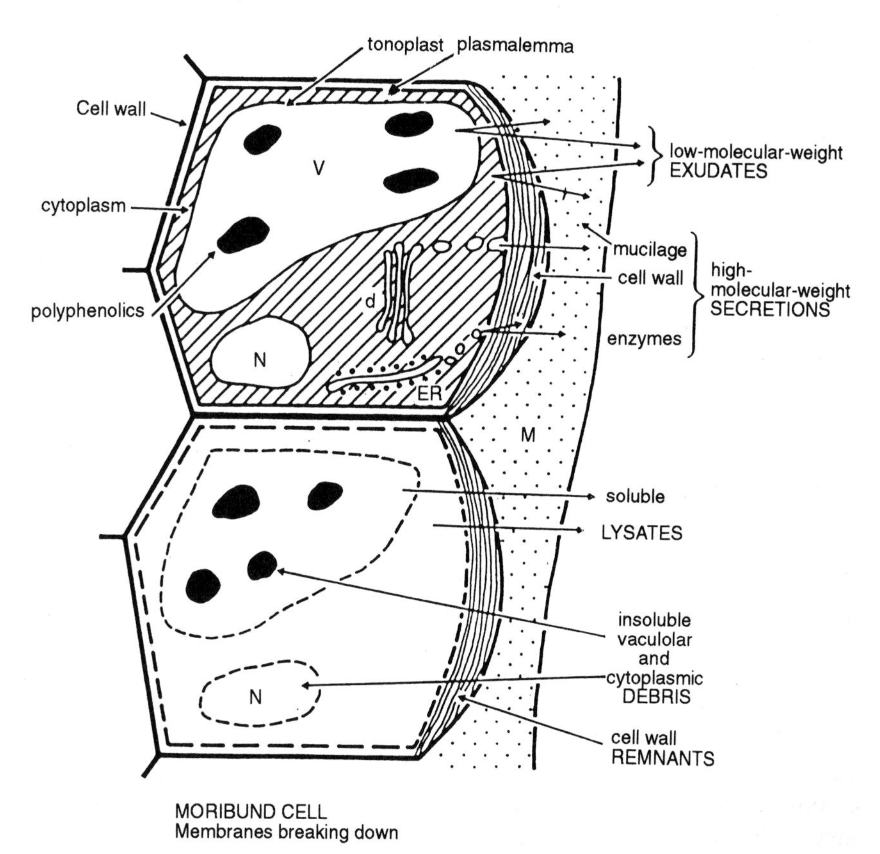

Figure 3.6 Diagram of root cells showing the sources of the major root-derived material found in soil (ER, endoplasmic reticulum; M, mucigel; N, nucleus; V, vacuole). After R.C. Foster and J.K. Martin (1981), In *Soil Biochemistry* Vol. 5, eds. E.A. Paul and J.N. Ladd, Marcel Dekker, New York, pp. 75–111.

Table 3.2 Carbon fluxes in the rhizosphere of a 35–50-year-old Douglas fir stand. After R. Fogel and G. Hunt (1979) *Canadian Journal of Forest Research* **9**, 245–256

Component	Flux (g m^{-2} per year)	Percentage of total
Total plant material	3032	100
Mycorrhizal sheath	610	20
Sclerotia and sporophores	232	8
Hyphae	699	23
Total fungus	1541	51

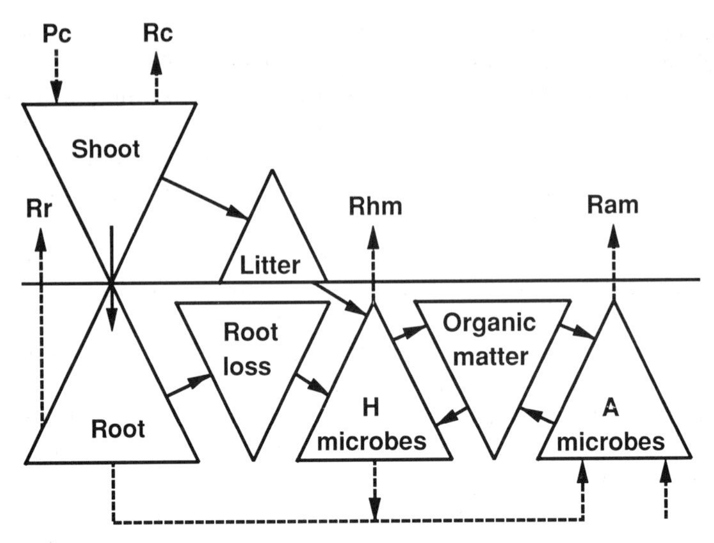

Figure 3.7 The transformations of C in an arable soil (Rc, respiration by the canopy; Pc, photosynthesis by the canopy; Rr, root respiration; Rhm, respiration by the heterotrophic (H) microorganisms; Ram, respiration by the autotrophic (A) microorganisms; solid lines, fluxes of organic C; broken lines, fluxes of CO_2. From M. Wood (1987) *Plant and Soil* **97**, 303–314.

The limited data available suggest that the major flux of CO_2 in the rhizosphere is due to microbial decomposition of root material. It has been estimated that 370–960 g m^{-2} root-derived material is lost from a crop of barley (*Hordeum vulgare*) growing in the UK.

3.5.1 *Amounts of substrates*

Bacteria require a supply of substrates for maintenance of their cells (for example repair of DNA) as well as for growth (increase in cell size and number), and the concepts developed from pure culture studies may be applied to bacteria in soil. The substrate requirements for a bacterial population can be expressed as

$$\mathrm{d}S/\mathrm{d}t = 1Y\ \mathrm{d}x/\mathrm{d}t + ax$$

where $\mathrm{d}S/\mathrm{d}t$ is the rate of input of substrate, Y is the growth yield of bacteria per unit substrate, $\mathrm{d}x/\mathrm{d}t$ is the growth rate, x is the biomass, and a is the specific maintenance rate. There is little information on maintenance rates or yield coefficients for bacteria in soil, although a growth yield coefficient of 0.35 has been assumed.

If a maintenance rate of 350 g g^{-1} per year (0.04 g g^{-1} h^{-1}) is assumed, then the bacterial population in the UK forest soil described in Table 2.4

would require 1300 g m^{-2} per year for maintenance alone. The estimated annual input of substrate as above-ground litter (545 g m^{-2}) and root-derived material (162 g m^{-2}) of 707 g m^{-2} is therefore apparently insufficient to support this bacterial population. Similarly, for a Canadian grassland soil the estimated annual maintenance requirement for a bacterial population of 55 g m^{-2} is 19 270 g m^{-2} compared to an estimated annual input of substrate of 500 g m^{-2}.

Calculations such as these lend support to the view that growth of the soil bacterial population is severely limited by substrate supply. When similar calculations are applied to the soil fungal population (measured as hyphal length by the agar film technique) in the UK woodland soil described in Table 2.4, the estimated maintenance requirements exceed the annual input of substrate. This reflects our lack of understanding of the nutrition and activity of the fungal population. The data for the US grassland soil shown in Table 2.5 assumed that only 10% of the fungal hyphae measured by the agar film technique were active.

One aspect of the metabolic state of microorganisms in soil has been exploited as a means of estimating soil microbial biomass. If it assumed that all microorganisms contain the same proportion of ATP, then the amount of ATP present in soil can be related to the amount of microbial biomass (assuming larger organisms are absent from the soil sample). ATP is extracted from soil using paraquat and tri-chloro-acetic acid, and measured by the amount of light produced when it reacts with luciferin-luciferase (obtained from firefly tails). A correction factor is used to allow for the efficiency of extraction of ATP from soil.

3.5.2 Survival strategies

The soil microbial population appears to maintain a high ATP content despite the limited substrate supply to these organisms. This supply may be supplemented by cryptic growth (utilisation of dead cells as substrates), metabolism of endogenous energy reserves (such as poly-hydroxybutyrate) or decomposition of more resistant fractions of the soil organic matter.

Soil may be considered as an oligotrophic environment, a concept developed for nutrient-poor aquatic ecosystems. The bacteria isolated from oligotrophic environments show relatively high growth rates at low concentrations of substrates and low maximum growth rates, and they can be isolated on low nutrient media. If it is assumed that substrates are evenly distributed in soil, and that all soil bacteria are equal in biomass, ATP content, metabolic activity and ability to compete for limited nutrients, then soil does appear to be an oligotrophic environment. Many soil bacteria appear unable to grow on high nutrient media; estimates of bacterial populations in soil using plate counts are considerably lower than those obtained by direct microscopy, but are increased by using low nutrient media. A low nutrient soil extract medium is recommended for the isolation of *Arthrobacter* sp. from soil. The concept of an oligotrophic soil bacterial population is similar in many respects to that of the autochthonous population (see chapter 2).

Competition between organisms for nutrients in soil is likely to lead to specialisation in terms of rate of growth and substrate utilisation. Rather than the whole of the soil bacterial population growing slowly and continuously (the oligotrophic or autochthonous population), part of the population may have periods of dormancy interspersed with periods of relatively rapid growth (the copiotrophic or zymogenous population). This may lead to a succession of organisms in response to an input of organic material in which the zymogenous population responds rapidly to the readily available substrates, followed by the slower-growing autochthonous population utilising less readily available substrates. Such a rigid distinction is unlikely to hold in soils where, for example, bacteria that are zymogenous under one set of conditions can be non-zymogenous under another.

A similar scheme has been proposed for fungi, in which the rapidly growing sugar fungi are the primary colonisers, utilising simple compounds, followed by the slower-growing cellulolytic and ligninolytic fungi. The decomposition products of the secondary colonisers may provide further substrates for the initial colonisers. This, together with seasonal fluctuations in supply of litter and root material produces temporal variation in substrate supply.

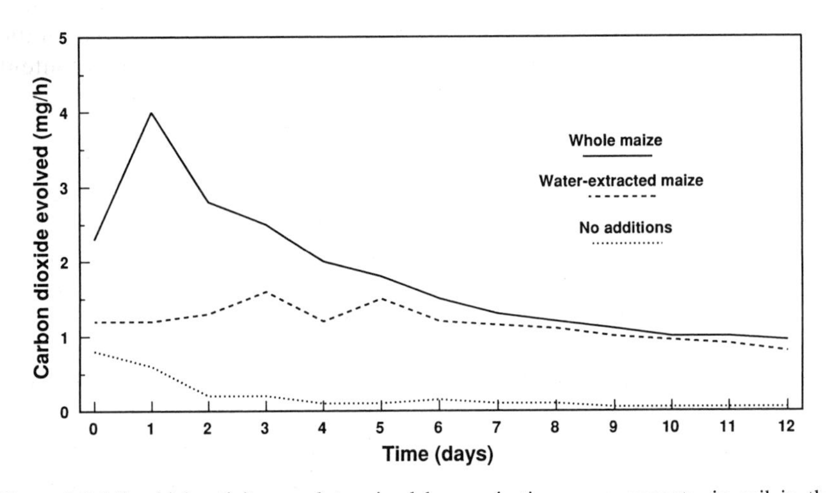

Figure 3.8 Microbial activity, as determined by respiration measurements, in soil in the laboratory following addition of maize stover either untreated or extracted with water. After A.S. Newman and A.G. Norman (1943) *Soil Science* **55**, 377–391.

These principles can be illustrated by considering what happens when fresh litter is added to soil. Figure 3.8 shows that following the addition of maize stover to a soil there is a rapid increase in biological activity, as indicated by respiration, which declines after a few days. This is due to the decomposition of readily available soluble organic compounds by the zymogenous component of the soil population; this can be eliminated by extracting the litter with water before addition to soil (Figure 3.8). The longer-term, slower rate of activity after about 7 days can be attributed to an autochthonous component which could decompose the more resistant parts of the litter.

There is also considerable spatial variability in substrate supply in soil. Litter and animal excreta are not uniformly distributed on the soil surface, and roots provide localised zones of high nutrient concentration. At a smaller scale bacterial cells provide substrate for organisms such as the predatory bacterium *Bdellovibrio* sp., and the cell walls of dead fungal hyphae are attacked by streptomycetes. Exudates from fungal spores may also provide high local concentrations of substrates. Nutrients may accumulate at interfaces such as the negatively-charged surfaces of clays and organic matter. It is therefore an over-simplification to consider the soil as a uniformly oligotrophic environment.

3.6 Substrate quality

The input of organic material into soil from living and dead organisms provides substrates for growth of organisms (detritivores), primarily microorganisms, and subsequently for the mesofauna, macrofauna and flora. However, the composition of this organic material reflects that of the organisms from which it is derived. It therefore varies greatly in its content of elements such as N and P, and also in the types of molecules it contains. For example, soluble organic materials lost from roots contain a wide range of compounds including sugars such as glucose and arabinose, amino acids such as glutamic acid and leucine, organic acids such as oxalic and propionic acids, nucleotides such as adenine and guanine, and enzymes including invertase and protease. Simple soluble molecules such as glutamic acid are readily assimilated by a wide range of organisms, whereas the ability to metabolise the complex and insoluble molecule lignin is restricted to a few fungi. Furthermore, the presence of certain compounds such as polyphenols may modify the availability of other compounds.

The quality of substrates can therefore be considered in terms of elemental composition and availability. Elemental composition is considered here, and availability is considered in the next chapter within the context of organic matter turnover and soil development.

3.6.1. Elemental composition

Table 3.3 shows the major elemental content of a range of organic materials. Because of the major technical difficulties in directly extracting microorganisms from soil the data for the bacterium and fungus are obtained from laboratory cultures of these organisms. *Escherichia coli* is not normally found in soil, but data are not available for common soil bacteria. All four materials are dominated by C which comprises nearly half the dry weight. The other major elements present are O, N, H, P and S. Transformation of these elements, which make up about 95% of the cell dry weight, will be linked with mineralisation of organic compounds. This is illustrated by the link between C and N mineralisation.

A bacterial cell in soil can obtain nutrients from either organic sources such as amino acids, or from inorganic sources such as ammonium ions, but it must assimilate elements in a ratio approximately equal to that required for the elemental composition of its own cell. For example, C and N must be assimilated in a ratio of approximately 5:1 (Table 3.3). If a bacterium is to utilise all of the C in a substrate such as maize shoots, with a C:N ratio of 31.4, then it must obtain extra N from inorganic sources to satisfy its own C:N ratio. The uptake and assimilation of inorganic nutrients by bacteria during the decomposition of organic materials in soil is termed immobilisation.

During decomposition of organic material by microorganisms some of the C in the substrate is released as CO_2 by respiration, therefore as less of the C is assimilated, less N is required to meet the required C:N ratio. The proportion of C assimilated from a substrate (i.e. yield) varies but a value of 0.35 has been used to estimate bacterial growth rates in soil (section 3.5.1). The new bacterial biomass eventually serves as a substrate with a relatively low C:N ratio for other functional groups such as protozoa and nematodes which have C:N ratios of less than 10:1. Release of some of the C during respiration by the assimilating organism will now lead to an excess of N relative to C in the substrate, and part of the N is released into the soil in the form of NH_4^+. The release of inorganic nutrients during the decomposition of organic materials is termed mineralisation.

Carbon released as CO_2 during the initial decomposition stage will also reduce the effective C:N ratio of the substrate; if the substrate has a high

Table 3.3 Elemental composition (% dry weight) of some organic materials. After D.S. Jenkinson and J.N. Ladd (1981)

Material	C	N	P	S	C:N
Bacterium (*Escherichia coli*)	50	15	3.2	1.1	3.3
Fungus (*Penicillium chrysogenum*)	44	3.4	0.6	0.4	12.9
Maize (*Zea mays*) shoots	44	1.4	0.2	0.2	31.4
Farmyard manure (cattle on straw)	37	2.8	0.5	0.7	13.2

initial C:N ratio, as in the example of maize shoots, this will reduce the effective C:N ratio of the substrate to the value which still remains higher than that of the assimilating organisms, resulting in immobilisation. The inefficient use of C by some organisms which release a large amount of CO_2 per unit substrate (i.e. have a low cell yield) will reduce further the effective C:N ratio of the substrate. The fixation of atmospheric N_2 by some bacteria (section 3.3.1) provides an alternative strategy for utilising substrates with a high C:N ratio.

Mineralisation is brought about by a succession of organisms in soil. The role of higher trophic levels in determining mineralisation is illustrated in Figure 3.9. The relatively low efficiency of C assimilation by nematodes (approximately 10%) together with their feeding on microbial biomass with a relatively low C:N ratio suggests an important and perhaps overlooked role for this group of organisms in the mineralisation of N.

Immobilisation can have important consequences for plants. For example, the incorporation of straw with a high C:N ratio into soil before planting a crop may lead to immobilisation of the mineral N which would otherwise be available to support plant growth, and the crop may become N deficient. However, immobilisation may be beneficial. For example, in temperate regions the cultivation of land in the autumn causes a flush of mineralisation from disturbed organic matter and the NO_3^- produced by oxidation of the NH_4^+ thus formed, may be leached from the soil and eventually enter water supplies (see chapter 5). Immobilisation of this

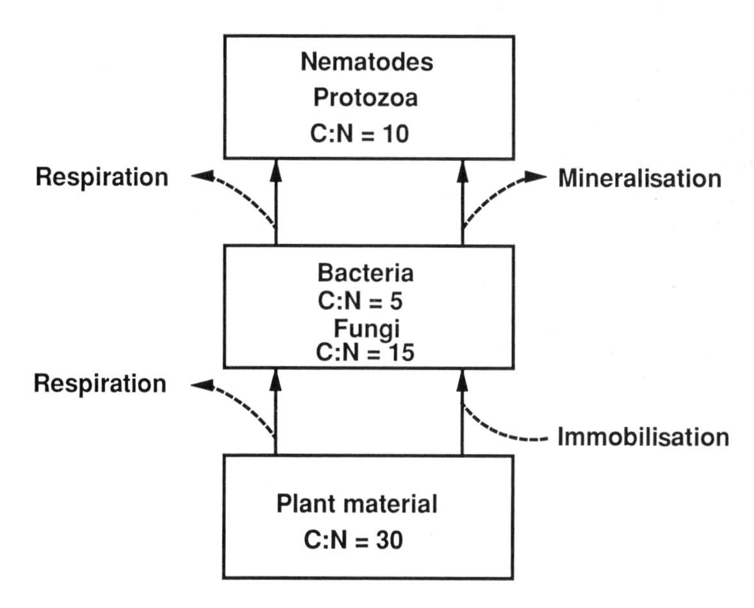

Figure 3.9 The role of different groups of soil organisms in the mineralisation and immobilisation of soil N.

mineral N into microbial biomass reduces this risk and allows a steady re-mineralisation of the N from the microbial biomass for crop growth during the following season.

These two examples illustrate the fine balance between mineralisation and immobilisation which often occurs in soil. Both processes may operate concurrently in a soil (this is referred to as mineralisation–immobilisation turnover or MIT) and the net result determines the availability of a particular nutrient to plants. There now appears to be a consensus that N mineralisation occurs if the N content of the substrate is greater than about 2%, and immobilisation occurs if the N content is less than about 2%.

Fungi generally have a higher C:N ratio than bacteria (Table 3.3) and may therefore immobilise less N per unit substrate assimilated than bacteria. However, such differences may be offset by the fungi having a higher efficiency of C assimilation (less C lost as CO_2). Animal manures with a lower C:N ratio are less likely to cause immobilisation than plant materials (Table 3.3). The residue of a legume such as medic (*Medicago littoralis*), with a C:N ratio of 15:1, is likely to mineralise N from the start of decomposition, differing in this respect from cereal or grass residues, which normally immobilise N during the early stages of decomposition.

3.7 Microbial biomass and nutrient cycling

The soil microbial population, more precisely their cellular constituents, can be viewed as a potential source of nutrients. Ecologists consider nutrient cycling in terms of food webs in which the interactions of different functional groups bring about the transfer of energy and nutrients. This approach relies upon an understanding of the roles of different groups of organisms in soil, and the complex nature of species interactions can lead to complex food webs.

Figure 3.10 shows a simple cycle, comprising three trophic levels, which forms the pattern of most nutrient cycles. The plants (primary consumers) are consumed by animals (herbivores) which excrete waste products and eventually die thereby providing substrates for the microbial biomass (detritivores). Dead plant material not consumed by animals is also consumed by the microbial population. The concept of microbial biomass is useful when considering the recycling of nutrients within the soil. This cycle can be described in terms of the sizes (pools) of the functional groups, for example the microbial biomass, and the rates of transfer (fluxes) of nutrients between pools.

The Broadbalk experiment at Rothamsted provides a relatively simple system which has been intensively studied. On one plot wheat (*Triticum aestivum*) has been grown continuously since 1843 without addition of fertiliser or manure, and the organic matter content of the soil has remained almost unchanged since 1881. The system is therefore at steady

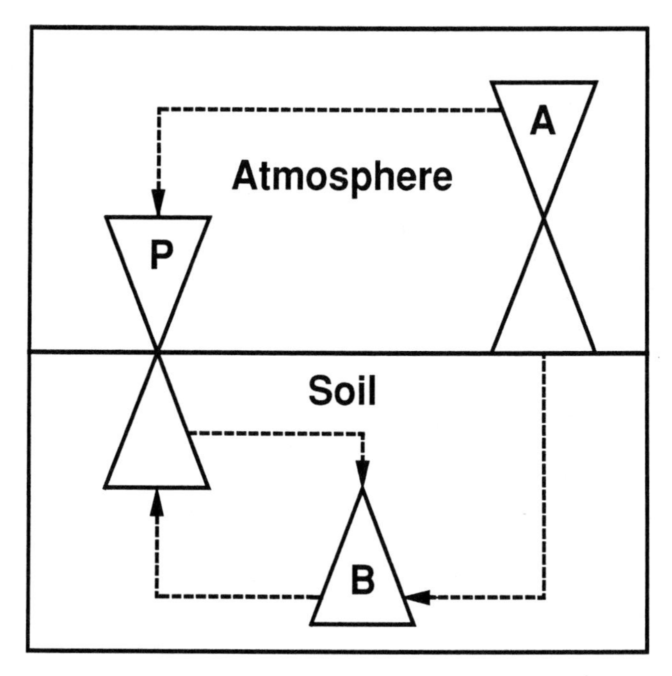

Figure 3.10 The major pools and flux pathways for nutrients in soil (A, animals; B, microbial biomass; P, plants).

state with no net loss or gain of nutrients. The amount of N and P contained in the microbial biomass (pool sizes) is 9.5 g m^{-2} and 1.1 g m^{-2}, respectively. Assuming the turnover time (see chapter 4) of the microbial biomass to be 2.5 years (as measured for biomass C) then the annual fluxes of N and P through the microbial biomass can be estimated to be 3.8 g m^{-2} and 0.5 g m^{-2}, respectively.

The annual removal of nutrients N and P at harvest from the unmanured plot at Broadbalk has been estimated to be 2.4 g m^{-2} and 0.5 g m^{-2}, respectively, indicating that the flux of nutrients through the microbial biomass makes an important contribution to nutrient supply to the crop. The origin of this input of N could be free-living N_2-fixing organisms; the P is probably derived from mineralisation or solubilisation of otherwise unavailable forms of P.

3.8 Summary

This concludes our cycle through some important biological processes in soil. It has not been exhaustive in its coverage, but has attempted to link the fundamental biochemistry of soil organisms to the transformations of elements which are of importance in the wider environment.

4 Soil formation and development

4.1 Introduction

The first three chapters of this book have considered the various components of soil, and some of the underlying biochemical processes which make soil one of the most fascinating living materials. Some scientists have been drawn to consider soil as a living tissue, with the soil solution acting like blood, transporting nutrients and waste products.

As with the study of any living organism therefore it is essential to study soil not in isolation, but within its own particular environment. Furthermore, it is important to realise that soil is a continuously developing material. This has been appreciated by pedologists, when considering the distinguishing features (e.g. horizons) of a soil in a particular landscape. Soil organisms play a major role in the formation and development of soils (pedogenesis), and this forms the subject of this chapter.

4.2 Pedogenesis

Soil can be described as a three-dimensional body within a landscape, comprising inorganic and organic material which changes with time. It is formed from weathered rock and mineral material, but the point at which this material becomes soil is not clearly defined.

A variety of factors contribute to the formation of a soil profile. A parent material of low base status produces an acid soil, whereas a material of higher base status gives rise to more alkaline soils. The shape of the land surface (relief or topography) is important; for example, waterlogging in low-lying poorly-drained areas causes localised reduced zones indicated by gleying or mottling of the soil (section 4.5.1). Climate can affect the rate and extent of leaching of bases, this being greater in wet tropical climates. Biological factors such as the type of vegetation are also important, for example, conifer trees produce acid leaf leachates which may mobilise some elements in soil and give rise to distinct eluviated horizons, as seen in podzolised soils.

A particular soil is therefore the net result of biological, topographical and climatic factors interacting with a particular parent material; the balance between these factors may alter with time. The various processes contributing to soil genesis can be considered as inputs, outputs, transfers and transformations (Figure 4.1). The resulting profile is the net effect of these processes operating on a particular parent material in a particular climate. These four processes can be illustrated by reference to the N cycle: inputs of N are often derived from N_2 fixation, outputs from denitrification, transfer of organic N is mediated by soil animals, and transformation of NH_4^+ to NO_3^- is carried out by nitrifying bacteria. The activities of man may influence pedogenesis by affecting any of these factors or processes.

Furthermore, soil development can be viewed as the net result of the processes of soil formation and soil erosion. This can be viewed as another example of the cycle of nature; material eroded from one site may be the starting material for soil formation at another site (Figure 4.2).

Geochemical evidence should also be taken into account when considering the long-term productivity of soils. It is reckoned that 30% of the chemical elements are essential for life; some of these such as Na and K are concentrated in the earth's crust, whereas others such as Ni and Cu are found in the deeper mantle. Soil organisms require elements from all parts of the earth's system; the greatest variety is likely to be found at sites on the earth's surface where convective mixing occurs.

Areas along the boundaries of tectonic plates such as the great mountain ranges of the European Alps and Himalayas, and areas such as Java, the Philippines and New Zealand are also some of the most fertile regions of the world. On the other hand, soils in stable continental regions which have been subject to weathering processes for maybe 10 million years (and in the case of the Amazon Basin have profiles 100 m thick) show depletion of trace elements such as Co and K.

4.3 Weathering of rock

The weathering of rock is one of the pre-requisites for soil formation. Exposed rock is often colonised by microorganisms, and even in the Antarctic dry valleys, possibly the harshest environment on earth, rocks are colonised by cyanobacteria. Microbial activity contributes to the chemical, physical and mineralogical changes which are involved in weathering. A range of processes such as acid hydrolysis, complex formation, oxidation and reduction, and exchange reactions cause solubilisation of minerals. Some elements may also become less soluble, for example by oxidation of Fe and Mn, by reduction of S compounds and by formation of carbonates.

The uptake of ions such as K^+ by microorganisms and plants can lead to the exchange of ions into solution from minerals. For example, in laboratory culture the fungus *Aspergillus fumigatus* can use micas as a source of K; as K is immobilised by the fungus, Na ions in solution exchange for K ions in the micas which results in the formation of

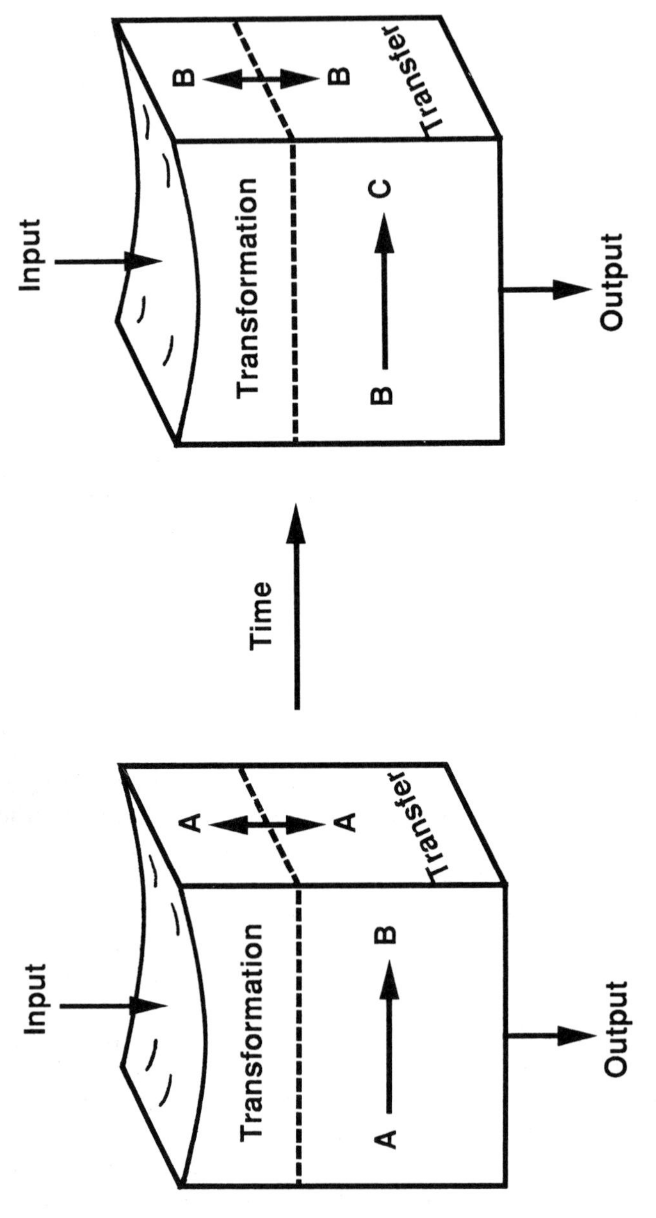

Figure 4.1 Hypothetical scheme for soil horizon differentiation. Based on R.W. Simonson (1959).

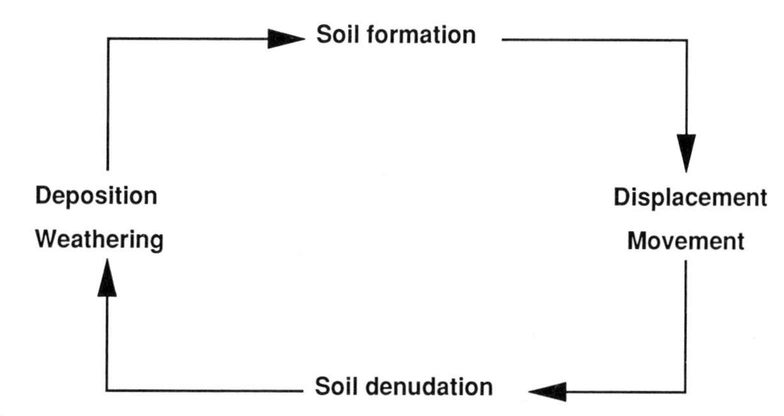

Figure 4.2 The cycle of soil formation and soil denudation forms the basis of soil development in many parts of the world.

vermiculite. Wheat plants (*Triticum aestivum*) and seedlings of a range of coniferous and deciduous trees growing in sand culture with K and Mg supplied as biotite, can convert biotite to vermiculite and kaolinite within 12 months. However, it is not clear from these studies whether the observed weathering was due to the roots themselves or to the associated rhizosphere microflora.

Physical forces such as those involved in the freezing of water in cracks and fissures in rocks are important in breaking rocks into smaller particles. Lichens, one of the first organisms to colonise exposed rock, may also play a part in this process. A thallus or hypha growing through a crack will exert pressure which may cause further disintegration. It has also been suggested that the gelatinous substances produced by this symbiotic organism when it dries on the rock surface may tear away mineral fragments in the same way that a layer of drying gelatin does on glass.

A decrease in particle size increases the specific area and hence the rate of solubilisation of minerals. There is little *in situ* evidence of a role for organisms in the solubilisation of minerals, but many laboratory experiments have shown that microorganisms can solubilise a range of elements and transform minerals. *Thiobacillus* spp. have been isolated from particular types of lesions in limestone. The characteristic scaling of limestone buildings (swelling, cracking and eventually falling away of the surface layer to leave a powdery zone) is linked to the conversion of $CaCO_3$ to $CaSO_4$ by these bacteria. Basic ferric sulphates are also produced by the action of sulphuric acid produced by *Thiobacillus ferrooxidans* on minerals and rocks.

Microorganisms have the potential to produce a wide variety of complexing or chelating agents that can solubilise elements such as Al, Fe,

Mn, Ni, Zn, and Ca. Both aliphatic acids, for example oxalic acid and 2-ketogluconic acid, and aromatic acids, for example salacinic acid, may alter the solubility of different elements. Many *Clostridium* sp. and *Bacillus* sp. produce volatile and semi-volatile organic acids such as acetic and butyric acids which solubilise Ca and K from granite sand. Complexing agents such as oxalic acid are more important in the solubilisation of Al and Fe which may be carried out by *Pseudomonas* sp. 2-ketogluconic acid has been identified during the microbial weathering of silicates and during the solubilisation of $CaCO_3$, montmorillonite and hydroxyapatite.

The difficulty is to obtain information on the reactions between specific chelating agents and mineral particles within soil under natural conditions, in which several processes may be operating concurrently. For example, in waterlogged soil the simultaneous production of organic acids by bacteria and the reduction of Fe and Mn may lead to the modification of biotite to vermiculite, and illite-vermiculite to montmorillonite.

The colonisation of rock by lichens presents a simpler system in which the environment is more uniform and the antibiotic properties of lichen acids, together with the otherwise unfavourable conditions, limit colonisation by other microorganisms. Lichens are symbiotic organisms formed by a cyanobacterium or green alga and a fungus, which may fix N_2, and produce large quantities of extracellular acids (comprising as much as 2–5% of the lichen dry weight). The rock lichen *Parmelia conspersa* produces salacinic acid which when added to ground granite and mica reacts within a few hours. The intimate contact between lichen hyphae and rock, together with the production of extracellular acids suggest a major role for these organisms in weathering of rock.

The ability of cyanobacteria and lichens to withstand drought and extreme temperatures, and to fix CO_2 and N_2, makes them well adapted for primary colonisation of rocks. Morphological features such as a reduced evaporative surface, thickened cortex and various amorphous and sometimes gelatinous external layers, together with pigmentation, assist in the protection of lichens against desiccation and excessive radiation from direct sunlight. Values for cyanobacterial biomass of 2–62 g m^{-2} rock have been reported for desert areas in the Middle East and North America.

Apart from a direct effect on the physical and chemical weathering of rocks, these organisms have an indirect effect in providing the first input of organic matter to the weathered rock. Humic acids formed from such organic material are active in weathering processes. For example, Fe and Al from minerals form soluble complexes with fulvic acid which allows leaching of these elements, as may occur during podzolisation. The reducing properties of fulvic acid and polyphenols may also increase the mobility of Fe as it is converted from Fe^{3+} to Fe^{2+}, and this may allow podzolisation to occur under aerobic conditions.

The organic matter produced by the primary colonisers allows growth of heterotrophic organisms and subsequent colonisation by plants. The N_2 fixed by lichens may be readily available to plants; experiments using ^{15}N-labelled N_2 showed that part of the atmospheric N_2 fixed by lichen crusts in desert grassland soils became available to plants growing in the crusts within 160 days. These later colonisers of rock and rock-derived material may then cause further weathering.

4.4 Organic matter

Organic matter constitutes the heart of soil, and as weathered rock accumulates organic matter it begins to show features more characteristic of soil. Organic matter influences many properties of soil such as fertility, structure and profile development. These effects are often inter-related, for example, the fertility of the soil will influence the types of plants which grow, and the type of plant may influence the development of the soil profile.

The accumulation of organic matter leads to horizon differentiation in most soils. The effect of organic matter on soil processes will depend not only upon the amount of organic matter but also on its rate of decomposition. In the early stages of development the input of organic matter is greater than the loss, and as time progresses the gains and losses move towards an equilibrium. In a few relatively rare cases, for example where the vegetation cover has remained the same for a long period of time, it can be assumed that the input of organic matter from the vegetation is constant during that period and steady state conditions exist. Few experiments have tested this assumption. However, if such conditions exist then the rate of decomposition or turnover of organic matter can be defined as:

$$\text{turnover time (years)} = \frac{\text{total soil organic matter}}{\text{annual input of organic matter}}$$

The decomposition of organic matter is a complex and poorly understood process which involves soil animals and microorganisms. It is influenced by the nature or quality of the organic matter, the climate, the soil conditions such as acidity, and especially man's actions (section 4.6).

4.4.1 Organic matter turnover

Table 4.1 shows the rates of input and of turnover of organic matter for soils from a range of ecosystems. Only the example of the wheat soil is known to be under steady state conditions; these data come from one of the long-term field experiments at Rothamsted Experimental Station where

Table 4.1 The turnover of organic carbon in soils from different ecosystems. After D.S. Jenkinson (1981)

Ecosystem	Net annual primary production ($g\ C\ m^{-2}$)	Annual carbon input to soil ($g\ m^{-2}$)	Carbon in soil ($g\ m^{-2}$)	Turnover time (years)
Temperate arable	260	120	2600	22
Temperate forest	710	240	7200	30
Subhumid savannah	140	50	1700	34
Tropical rain forest	950	490	4400	9

levels of organic matter have remained constant in soils growing wheat continuously since 1843. Organic matter turnover is fastest in tropical rain forests where the higher temperature and humidity and the action of termites accelerates decomposition.

The turnover times given in Table 4.1 are average values for all of the soil organic matter. However organic matter is far from homogeneous. In the examples used in chapter 3 (section 3.6.1) to illustrate the effect of the elemental composition in determining the fate of a substrate, it was assumed that all of the C and N in the substrate was immediately available to the bacteria. However, studies using tracers such as ^{14}C and ^{15}N to follow the decomposition of organic materials in soil indicate otherwise.

Figure 4.3 shows the decomposition of ryegrass (*Lolium perenne*) labelled with ^{14}C. Decomposition appears to follow two distinct phases: a relatively short and rapid initial phase, followed by a longer slower phase. During the first year about two-thirds of the carbon is lost from the plant material. Some of the breakdown products from this material are converted into more resistant humic material and this new material, together with more resistant plant material is decomposed during the second phase. Clearly a fraction of the C in the substrate remains unavailable to the microbial biomass for a considerable time. Decomposition of ^{14}C, ^{15}N-labelled medic (*Medicago littoralis*) in South Australian soils followed a similar pattern. More than 50% of the medic ^{14}C had disappeared after 4 weeks and only 15–20% of ^{14}C and 45–50% of ^{15}N remained as organic residues after 4 years. Data such as these can be used to develop simulation models of organic matter dynamics.

It has long been known that the chemical composition of plant litter influences its rate of decomposition. Early workers such as Waksman suggested that the content of water-soluble compounds, cellulose and hemicellulose, N complexes and lignin may all be involved. Over the subsequent years these ideas have been tested, but no great advance made in our understanding of the underlying processes.

Data for forest litter in Manipur State, India, indicated that the initial

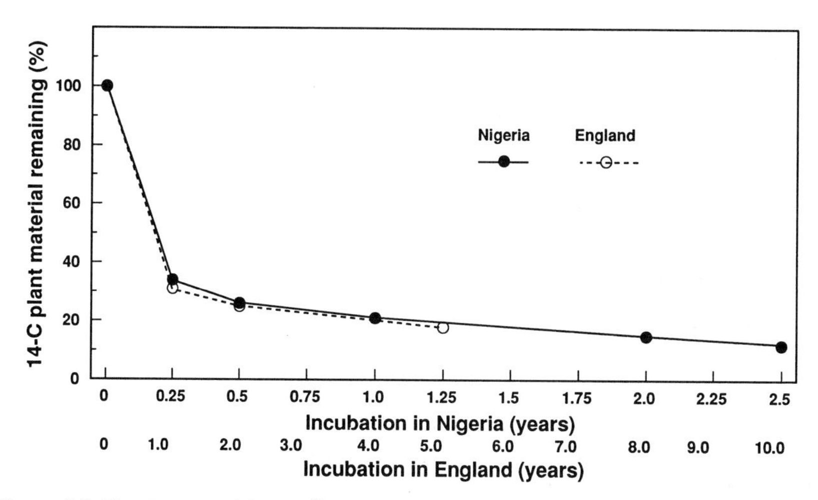

Figure 4.3 The decomposition of ^{14}C-labelled ryegrass (*Lolium perenne*) in the field under temperate and tropical conditions. After D.S. Jenkinson (1981); A. Ayanaba *et al.* (1976) *Soil Biology and Biochemistry* **8**, 519–525.

lignin content may be more important than initial N content in determining the rate of decomposition. Data for leaf litter of two tropical forage legumes in a yellow podzolic soil showed that the polyphenol content was the most important factor involved. Other data indicate that the ratio of polyphenol content to N content, rather than N content or polyphenol content alone, provides the best indication of rate of decomposition of tropical legumes in an acid soil.

In mature temperate and tropical forest soils 92–99% of the above-ground biomass is in the form of wood and a high proportion of the root system of trees is also likely to be wood. Although the input of wood as a substrate for soil organisms is often under-estimated, woody tissues such as branches are conserved by plants, whereas photosynthetic tissues such as leaves are not. Wood is also a low quality resource due to extensive lignification, low content of soluble sugars and minerals, and often high content of allelopathic compounds (see chapter 6), protection by bark, and the small external area to volume ratio associated with the often massive size of the substrate units (an oak tree may be 50 m^3).

Mathematical models have been developed to simulate the turnover of organic matter in soils; many models focus on the early stages of decomposition over periods of days, weeks or months. There are fewer models which are aimed at simulating longer-term changes over years or centuries (Table 4.2). These models vary in their complexity, but nearly all of the models partition organic matter into different fractions and then ascribe a turnover time to each particular fraction (Figure 4.4).

Table 4.2 Three models used to simulate long-term changes in organic matter in soils

Model	Reference
Rothamsted Carbon Model	D.S. Jenkinson (1990) *Philosophical Transactions of the Royal Society London B* **329**, 361–368.
CENTURY	W.J. Parton *et al.* (1994) in *The Biological Management of Tropical Soil Fertility*, eds. P.L. Woomer and M.J. Swift, John Wiley and Sons, Chichester, pp. 171–188.
SCUAF	A. Young (1989) *Agroforestry for Soil Conservation*, C.A.B International, Wallingford.

Turnover times for these fractions reflect our current lack of understanding of decomposition processes in soils. Examples of values used include 1.7 years for biomass and 2400 years for the most resistant fraction. These models are usually tuned and tested using data from decomposition of ^{14}C-labelled organic matter in the field, or from long-term agronomic experiments.

The material in a compartment is assumed to decay by first-order kinetics as follows:

$$dC/dT = A - kC$$

where dC/dT is the rate of change in the amount of material in the particular compartment, A is the annual addition of material to that compartment, and k is the rate constant (the fraction of the material decomposed each year; note that the turnover time is equal to $1/k$). It is also assumed that the rate constant of the material is a property of the organic material itself, and is not influenced by the composition of the microbial population.

In most of these models the rate constants for the various compartments are multiplied by one or more rate modifiers which acknowledge the effect of factors such as temperature, soil moisture, clay content and cultivation. The CENTURY model was initially designed and validated for the US Great Plains grasslands, and has now been extended by the use of data sets for a wider range of environmental conditions. SCUAF was designed to model the special features of agroforestry systems. The Rothamsted Carbon Model was one of the earliest models of soil organic matter dynamics, and has the unique advantage being validated against data from 150 years of continuous field experiments in England.

Table 4.3 shows the sizes of the estimated organic matter fractions for a Canadian pasture soil. It has been estimated that 50–60% of soil organic matter is protected from decomposition due to its association with soil solids. For example, amino acids adsorbed to clay particles are degraded

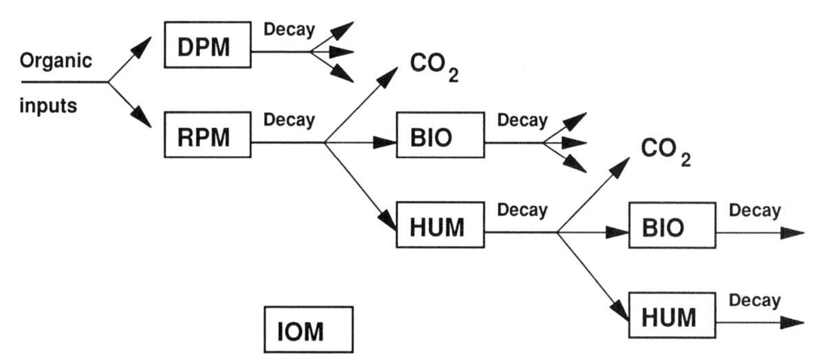

Figure 4.4 The basic structure of the Rothamsted Carbon Model (DPM, decomposable plant material; RPM, resistant plant material; BIO, microbial biomass; HUM, humus; IOM, inert organic material). After D.S. Jenkinson (1990) *Philosophical Transactions of the Royal Society London B* **329**, 361–368.

Table 4.3 Major sources and estimated amounts of organic substrates in a Canadian pasture soil. After E.A. Paul and R.P. Voroney (1980), in *Contemporary Microbial Ecology*, eds. D.C. Ellwood *et al.*, Academic Press, pp. 215–237

Source	Amount (g C m^{-2})
Litter	100
Root	1030
Microbial biomass	125
Organic matter	
Decomposable (not protected)	440
Decomposable (protected)	4410
Recalcitrant (not protected)	1700
Recalcitrant (protected)	1700

much more slowly than free amino acids. The proportion of those parts of perennial plants that becomes available to the microbial biomass annually is not clear, although a rate constant of 0.08 per day has been used in a simulation model of C turnover in this soil.

4.4.2 Effects of seasonal change

The major climatic features of the different ecosystems were introduced in chapter 1. Tropical regions in particular are characterised by a wide variety of different climates ranging from arid regions to those with pronounced wet and dry seasons, to those with a continuously wet climate. The effects of these different climatic conditions are likely to have a great impact on biological activity including litter decomposition and nutrient cycling in tropical soils. Indeed the variation in climatic conditions within the tropics may be greater than any differences between temperature and tropical regions, which means that much of the research work carried out in temperate soils may be relevant to tropical soils. This information is also valuable when considering possible effects of global climate change (see chapter 6).

A major consistent difference between temperate and tropical regions is the temperature; on average the tropics are 15°C warmer than temperate regions. This causes an increase in the rates of chemical and biochemical reactions. Where moisture is not limiting, for example in humid forests, the rates of primary production and decomposition are therefore several times more rapid than in temperate regions.

This is illustrated by data shown earlier (Figure 4.3) on the decomposition of ^{14}C labelled plant material in the field in Nigeria and in the UK. The mean annual temperature at the site in Nigeria was 26.1°C, and at the site in the UK was 8.9°C, i.e. a mean difference of 17.2°C, and the rate of decomposition in Nigeria was four times faster than the rate in the UK. This is in reasonable agreement with the van 't Hoff rule which states that an increase in temperature of 10°C leads to a doubling in the rate of reaction.

The effect of temperature on rates of decomposition in tropical soils may be modified by other factors such as litter quality (section 4.4.1) and soil properties such as mineral composition. For example soils derived from volcanic ash contain large amounts of allophane. This mineral forms complexes with soil organic material and microbial products which slows down the rate of decomposition of organic matter. It may be that the microbial biomass is also stabilised in these Ando soils. Studies on such soils from Chile showed that they were dominated by Steptomycetes and fungi.

Our understanding of the effects of well-defined wet and dry seasons on biological activity in tropical soils has its origins in earlier studies on partial sterilisation of soils. Early work in the UK by Russell and the USA by Waksman, on the use of fumigants to partially sterilise soil in order to control soil-borne plant pathogens, had shown that partial sterilisation caused a short-term increase in soil respiration compared to a non-fumigated soil. A similar response was observed for a Kenyan highland soil subjected to alternate periods of drying and re-wetting (Figure 4.5); drying increased the amount of carbon and nitrogen subsequently mineralised upon re-wetting. In the 1950s Birch concluded that in the field maximum nitrate production therefore occurs at the start of the rainy season, and that crops can best benefit from this flush of mineralisation by being planted before the rains start (dry planting).

The mechanism responsible for this flush of mineralisation following drying and re-wetting was established following experiments at Rothamsted

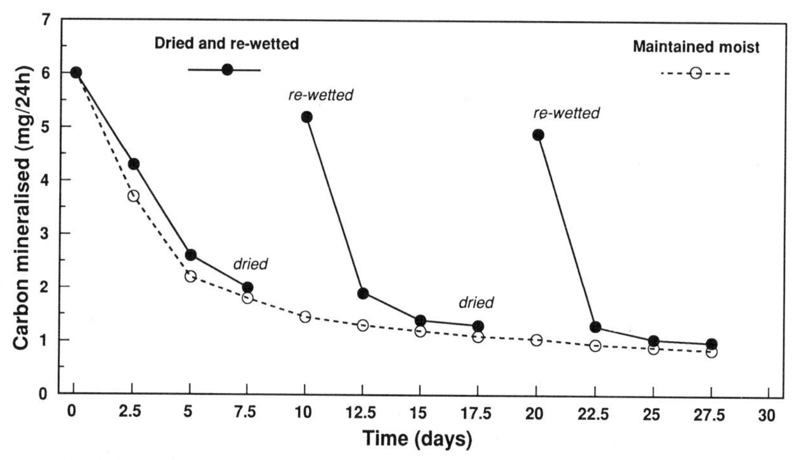

Figure 4.5 Effect of drying and re-wetting a moist Kenyan soil on C mineralisation as determined by respiration. After H.F. Birch (1958) *Plant and Soil* **10**, 9–31.

Experimental Station on partial sterilisation of soil. From these studies it was concluded that partial or complete sterilisation of soil caused an increase in the rate of mineralisation of the microbial biomass by killing or damaging microorganisms which, therefore, mineralised more rapidly than undamaged ones. It can be concluded therefore that the flush of nitrate produced in tropical soils at the start of the rainy season is due to the action of surviving bacteria and actinomycetes decomposing their recently killed comrades.

It is worth noting here that these studies led directly to the development of the chloroform fumigation technique which has become one of the standard methods for measuring the amount of microbial biomass in soils. If a soil sample is fumigated with chloroform most of the microbial population is killed. The amount of C, N, P and S released from this biomass following lysis can be measured and related to the amount of living microbial biomass originally present in the soil.

There are two main variations on this basic method, fumigation–incubation and fumigation–extraction. In the fumigation–incubation method the fumigated soil, after the chloroform has been removed, is inoculated with fresh soil. The colonising microbial population decomposes the killed microbial biomass and the amount of CO_2 produced over a period of 10 days is used to estimate the amount of C in the original microbial biomass. An unfumigated control is used to allow for decomposition of non-living soil organic matter by the colonising population. Not all of the killed biomass is decomposed by the colonising population and a correction factor (k_C, K_N, etc.) has been obtained either by comparing the data obtained from this method with estimates of biomass obtained by direct counts (a laborious and highly skilled method unsuitable for routine biomass measurements), or by measuring the recovery of introduced dead microbes.

In the fumigation–extraction method the soil is extracted immediately after fumigation and the amount of organic C or N in the extract is determined. The amount of amino-N in the extract may also be measured using the ninhydrin reaction and used to estimate the amount of biomass N in the soil. Such techniques have provided useful information on the effects of management practices on fluxes of nutrients (see chapter 3) and organic matter turnover (section 4.4.1).

4.4.3 Effects of soil conditions

Mention has already been made of the effects of certain soil conditions on litter decomposition. However, there are major chemical features of some soils which may have a marked effect on these processes.

Acid soils are found widely all around the world; associated with this acidity in soils are low amounts of available Ca, Mg and P, and high concentrations of available Mn and Al. Al has been identified as a major factor associated with poor plant growth in acid soils, but complex interactions occur between all of the soil acidity factors in determining plant growth.

The effects of soil acidity on litter decomposition have not received much attention. It has been suggested that plants growing in acid soils contain more polyphenols than those growing in neutral soils, and certain plants such as tea which grows well in acid soils may contain up to 20000 $\mu g\ g^{-1}$ Al. The implications of these changes in litter composition for subsequent decomposition in acid soils are not clear. Soil acidity is considered in more detail in chapter 5.

Salinity is another common feature of soils in the tropics. Soluble salts accumulate in the surface of soils under hot dry conditions when the groundwater comes within a few metres of the soil surface, as happens following forest clearance. Saline soils may also be produced by poor irrigation practice. These salts may exert a general osmotic effect and also a specific ion effect, particularly Na and Cl ions. Both plants and microorganisms have evolved mechanisms to tolerate saline soil conditions. Salinity is also considered in more detail in chapter 5.

Streptomyces spp. are highly tolerant of salinity and are frequently found in saline soils. Studies on alkaline alluvial soils in Iraq between the Tigris and Euphrates rivers near to the ruins of the ancient city of Babylon which had been cultivated without fertiliser for over 400 years revealed a significant proportion of the bacterial population to comprise actinomycetes.

4.4.4 *Role of earthworms*

Soil animals play a major role in decomposition of organic matter. Probably the best known example is that of the earthworm. This was the subject of Darwin's final book *The Formation of Vegetable Mould Through the Action of Worms, with Observations of Their Habits* published in 1881, which marked the culmination of over 40 years interest in these animals. Darwin's studies led him to the conclusion that 'Worms have played a more important role in the history of the world than most persons would at first suppose'. He observed that worms select their food which they then shred and partially digest and mix with earth to form dark, rich humus which encourages nitrification. He suggested that earthworm burrows allow air to penetrate deeper into the ground and allow the downward passage of roots into soil. Subsequent investigations have confirmed much of Darwin's work.

The formation of burrows is a feature of earthworm activity, although not all species have burrows; it is usually only those that go deep into soil, for example *Lumbricus terrestris*. Burrows are formed by worms eating their way through soil and pushing through cracks. They are 3–12 mm in diameter with smooth walls cemented with mucous secretions and ejected soil. The mucous may act as a substrate for growth by fungi. Smaller, shallow-working worms have no permanent burrows.

Burrowing species produce casts on the soil surface near burrow exits. There are many different forms of casts, and they are often typical of the species that produced them. European casts usually weigh less than 100 g, but the giant *Notoscolex* sp. found in Burma produces large tower-shaped casts up to 25 cm high and 4 cm in diameter weighing up to 1.6 kg. Darwin estimated the annual production of earthworm casts in English pastures to be 1870–4030 g m^{-2}. More casts are found in pasture soils than in arable soils, and this may reflect differences in organic matter input and earthworm populations. Larger amounts of cast material are reported for tropical soils, for example 5000 g m^{-2} in Ghana. The amount of soil passing through the animals may be larger than these values because some worms void their casts underground.

Earthworms generally prefer soils with near neutral pH values. The absence of worms in acid soils leads to the accumulation of a thick mat of slowly decaying organic matter at the soil surface characteristic of soils with mor humus (section 4.5.4). Soil crumb structure may also be poorer when worms are absent. The importance of earthworms lies in their role in fragmenting plant material, making it more readily decomposed by soil microorganisms, and in mixing this organic material with the soil. A few common species such as *Lumbricus terrestris* appear responsible for the fragmentation of most of the litter in woodlands of the temperate zone. *Lumbricus terrestris* can remove 90% of the autumn leaf fall in an apple orchard during winter, accounting for 120 g dry matter per m^2.

Different species of earthworms have different food preferences, for example *Lumbricus rubellus* is a litter feeder, whereas *Allolobophora caliginosa* consumes partially decomposed organic matter. *Lumbricus terrestris* pulls leaves and other plant material into the mouth of the burrow before feeding on it, thereby plugging the burrow. It carefully selects food material and pulls leaves into the burrow, usually by the tip of the laminae leaving the unpalatable petioles protruding from the burrow.

The ability of most species to distinguish between different kinds of plant litter may be due to differences in plant mineral content or alkaloid content. Weathering of many types of leaves is required before they are palatable to worms; this may allow the leaching of undesirable poly-phenolic compounds. Worms prefer moist litter to dry litter and they can consume 100–300 mg of plant material per gram of body weight per day. They pass a mixture of organic and inorganic material through their intestines when feeding and burrowing. The intestinal microflora of the worms may assist in humification, but the role of earthworms in decomposition as opposed to fragmentation of plant material is not clear.

4.4.5 Role of termites

Termites are often the dominant members of the soil fauna in tropical, subtropical, semi-arid and some warm-temperate regions of the world. A

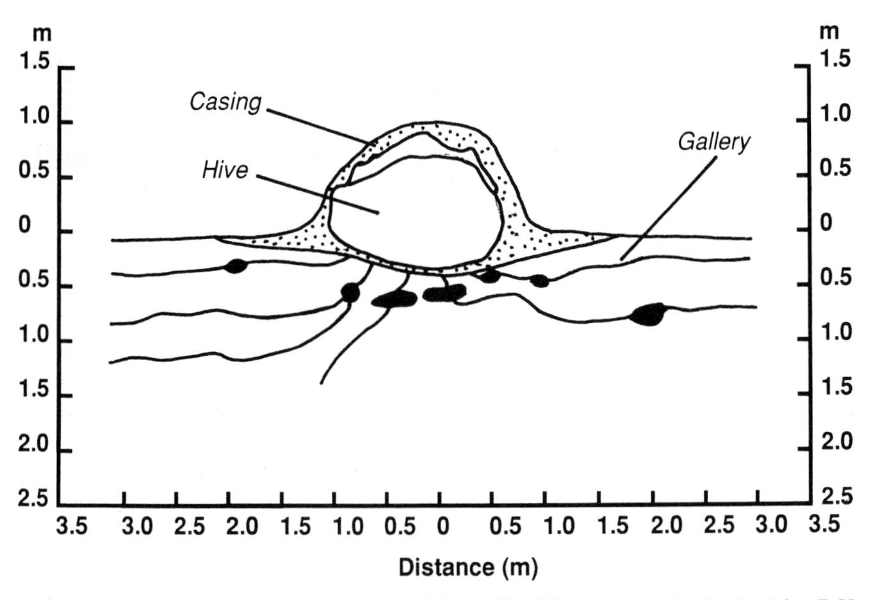

Figure 4.6 The structure of a termite mound formed by *Macrotermes nigeriensis*. After P.H. Nye (1955) *Journal of Soil Science* **6**, 73–83.

colony of termites forms a characteristic mound (Figure 4.6) often when forest has been cleared or thinned for agricultural use, but also in forests, sometimes covering trees and shrubs.

Mounds may be built rapidly, 60 cm in 1 month, and a single mound may contain 2.5 tons of earth. A coverage of 1–10 mounds per hectare have been observed in West Africa. The central part of the mound (the hive) has a complex system of galleries leading to food sources. Galleries radiate from the floor of the nest to about 90 cm depth and extend for about 3 m on either side of the mound. Galleries open out at regular intervals into chambers up to 90 cm wide. As food supplies are reduced galleries are abandoned and new ones exploited. A colony may last for 10–80 years and the mound is only slowly decomposed after being abandoned.

The colony is virtually a closed system in which unhealthy and dead individuals are consumed, and the only losses are due to predation and annual flights of secondary reproductives (alates). The release of alates, which are rich in fat and protein, over 1–2 h at favourable times of the year may represent up to 60% of the colony biomass. Predation on colonies and foraging parties is often by specialised predators, for example in Nigeria parties of up to 300 ants may capture 1000 termites.

Termites feed on all kinds of plant material (wood, grass and roots) both living and decomposing. They are particularly adept at consuming plant material before the material is attacked by saprophytic microorganisms. However, as most work on organic matter decomposition has been concentrated in cool temperate regions, the role of these organisms has not been fully appreciated.

A few species feed on living plants, preferring dry parts of grasses to fresh green tissue. Dead plant material is often attacked before it falls to the ground. Many termites, for example *Macrotermes* sp., eat dead material such as bark on living plants. Fresh woody litter is readily consumed. Termites also eat decomposing litter and dung, and there is sometimes a relationship between the stage of decay of the organic matter and the species feeding. This may involve effects of fungi on the material rendering it more readily available or palatable.

Many species consume soil rich in organic matter; these species show a morphological specialisation of the mandibles (mouthparts) which distinguish them clearly from those species feeding on decomposing organic matter. There is further specialisation in some species, for example *Hospitaltermes* sp. found in Malaysia feeds on lichens. Practices such as coprophagy (eating faeces), necrophagy (eating the dead) and cannibalism (eating the living) within the termite colony lead to efficient recycling and utilisation of nutrients.

The influence of termites is during the initial stages of decomposition in the savannah, but is during the later stages of decomposition in the rain forest. Most of the energy for termites comes from the breakdown of complex plant polysaccharides. Lignin is degraded and cellulases are produced in the gut, but, as with earthworms, it is not clear whether these enzymes are produced by the animal or by the intestinal microflora. The efficiency of assimilation is 54–93%, and the intense degradation of plant material gives rise to faeces with distinct chemical properties. Faeces have a high C:N ratio (107:21) and, particularly those produced by the wood feeders, have a low carbohydrate content and a high lignin content. Termites are unusual in that they use their faeces for constructing part of the nest system. However, the practice of coprophagy means that excreta may pass through the alimentary system of several individuals before being used.

The Macrotermitinae are a special case because all of their excreta are used to construct fungus combs. These termites are unable to decompose plant structural polysaccharides, but the fungus combs support a pure culture of *Termitomyces* sp. which degrades these compounds, and the products of their activity are ingested by the termites in the old portions of the comb. Combs are replaced by new ones every 5–8 weeks. The termites also eat white fungus spherules, rich in N, on the surface of the comb. Metabolism of C by the fungus leads to a reduction in the C:N ratio of the combs. Termites therefore consume more food than is required for their own maintenance. Also, some Macrotermitinae do not consume all of their food directly, instead they finely macerate the material and store it in heaps to allow conditioning.

An attempt to quantify the role of termites in decomposition is presented in Table 4.4 for savannah woodland in Nigeria (soil feeders are excluded). The total annual estimated consumption of 168 g m^{-2} represents a significant fraction of the estimated annual grass production of 200–300 g m^{-2} plus yearly litter fall from trees of 200–400 g m^{-2}. This rate of consumption assumes a greater significance when the loss of grass, leaf and woody material during bush fires is considered.

4.4.6 Food webs

Although many basidiomycete fungi possess the ability to decompose wood completely, under natural conditions decomposition occurs as a result of interactions between animals and fungi. The initial 60% of weight loss of wood in a temperate deciduous forest, which occurs mainly as death and decay in the canopy and branch fall, is due to fungi. The final 40% decomposition is due to the cooperative action of wood-boring animals and fungi. The fungal contribution may involve enzymatic softening of the wood, destruction of allelopathic compounds, and the production of attractants. Fungi may also improve the nutritional quality of the wood by reducing the C:N ratio. For example, in order to gain 10 μg of N an animal would need to consume 5.3 mg of undecayed wood, but only 1.9 mg of

Table 4.4 Role of termites in decomposition of wood, grass and fresh litter in Southern Guinea savannah woodland. After T.G. Wood (1976)

Measurement (dry weight basis)	Macrotermitinae	Others	Total
Number per m^2	1928	1540	3468
Biomass g m^{-2}	1.5	0.9	2.4
Annual consumption g m^{-2})	129.2	38.3	167.5
Annual returns (g m^{-2}) via			
Faeces	0	12.3	12.3
Alates and neuters	9.0	2.7	11.7

wood which had already lost 60% of its weight due to fungal colonisation. A considerable fraction of the nutrients in wood consumed by animals may be in the form of fungal mycelium. Finally, the penetration of bark by animals such as wood-borers and woodpeckers may allow colonisation by a secondary fungal population.

Termites are involved in decomposition of wood in tropical soils, and may attack wood before it falls to the ground. Conditioning of substrates by microbes prior to consumption by these animals is also important.

Studies on protozoa and nematodes in microcosms (simple, defined, laboratory-based soil communities) indicate that the interaction of these organisms with bacteria may be important in enhancing mineralisation of nutrients from organic matter, and in some cases increasing uptake of nutrients by plants. The grazing of bacteria by protozoa will be greatest where bacterial growth is occurring most rapidly, for example, in the rhizosphere and on freshly added organic substrates such as manure. In the rhizosphere there is likely to be adequate C for growth but a limited supply of N; grazing protozoa may release, as ammonia, approximately one-third of the N immobilised in bacterial biomass close to the roots. It has been shown in microcosms that the presence of protozoa increases mineralisation and uptake of N by plants.

There is some evidence of interactions between mesofauna and bacteria in field soils. After a dry soil is re-wetted during rainfall the numbers of bacteria increase, and this is usually followed by a rapid decline in numbers. Protozoa have been implicated in this decline in bacteria, which is unlikely to be due to either autolysis of cells (conditions should be favourable for bacteria), or a major bacteriophage (virus) attack. Furthermore, generation times for nematodes (also potential predators) are too long to allow a rapid increase in grazing pressure.

In the humus layer of a Swedish forest soil (pH 3.5–4.0), where most of the feeder roots were located, an increase in bacteria was observed 2 days after rainfall, followed by an increase in the population of naked amoebae after 5 days. Of the subsequent decrease in the bacterial population, 60%

could be accounted for by the increase in protozoa. Also, data from Sweden for a 120-year-old pine forest indicated that the fauna (1–7 g m^{-2}), although only a small proportion of the total biomass compared to bacteria (39 g m^{-2}) and fungi (120 g m^{-2}), consumed 30–60% of the microbial population in the litter and humus horizons, which accounted for 10–49% of the annual total N mineralised (2.8 g m^{-2}).

More detailed information on the flow of energy and nutrients between functional groups of organisms has been obtained from studies of a shortgrass prairie soil in Colorado, USA. The dominant plant is *Bouteloua gracilis*, and the biomass of the main functional group is given in Table 2.5. Using a combination of field data on population sizes and assumptions about C content, C:N ratios, death rates and assimilation efficiencies of the different groups, an annual budget for N was calculated based on the nutrient pathways shown in Figure 4.7.

These data show that the greatest amount of N is mineralised by bacteria (4.5 g m^{-2} per year) followed by the fauna (2.9 g m^{-2} per year). Amoebae and bacterivorous nematodes account for 83% of the mineralisation by the fauna. Very little of the N flows to the higher trophic levels of the mites and collembola. Values for direct mineralisation by fauna do not reflect their importance in this process because they return a large amount of N as labile substrates which are rapidly mineralised by bacteria. The role of the fauna is therefore to accelerate the recycling of nutrients within the ecosystem.

It has been suggested that a greater variety of species in a community leads to reduced fluctuations in population densities, and the greater variety of channels in a food web stabilises the flow of nutrients through that web. Figure 4.7, which represents the food web in a grassland soil, shows that compartmentalisation of food channels occurs at the base of the food web where the bacterial and fungal channels are quite separate, and the only common consumers, predatory nematodes and mites, are several trophic links away. There is also temporal compartmentalisation due to the seasonal pattern of substrate availability, and habitat compartmentalisation; the bacteria-based channels, involving bacteria, protozoa and nematodes all require a film of moisture for growth, whereas the fungi-based channels which involve mainly fungi, mites and collembola, do not require a continuous film of water. More data are required from a range of ecosystems to test fully the hypothesis that an increase in species diversity increases the stability of an ecosystem, when associated with an increase in the number of channels of energy flow. This is discussed further in chapter 6.

4.5 Profile development

The development of a soil profile may be influenced by a variety of biological factors ranging from the development of soil structure by

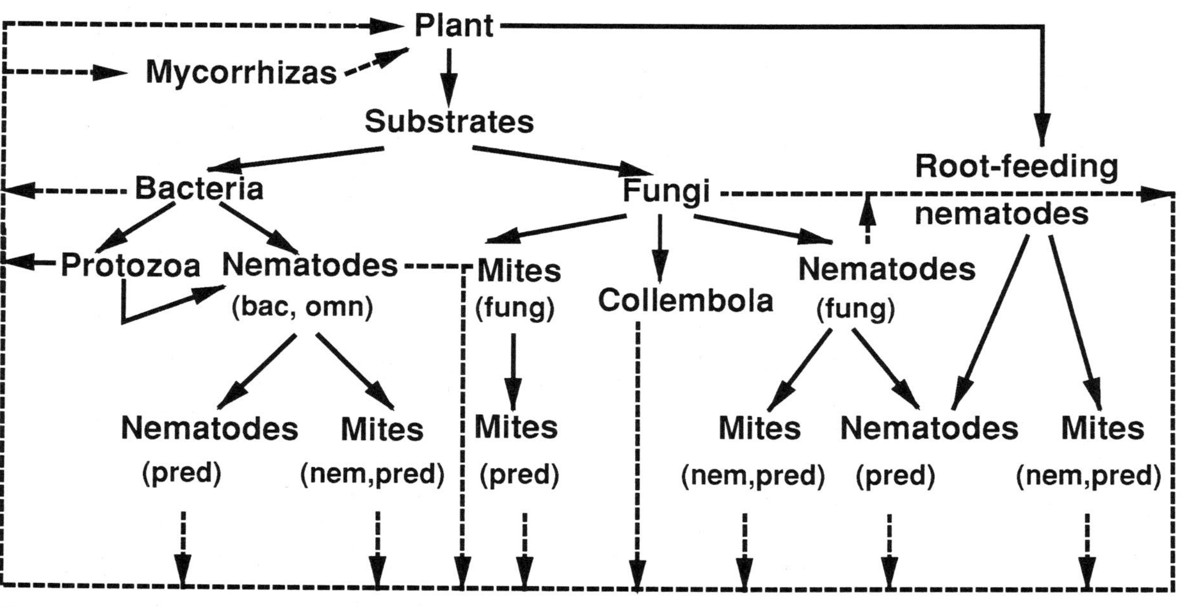

Figure 4.7 Flux pathways for N in a shortgrass prairie in the USA (bac, bacterivorous; fung, fungivorous; omn, omnivorous; pred, predatory; nem, nematophagous; solid lines, organic N; broken lines, inorganic N). Fluxes from all organisms to the substrate pool (death) have been omitted. Based on H.W. Hunt *et al.* (1987) *Biology and Fertility of Soils* **3**, 57–68.

microorganisms to the effects of vegetation in influencing the extent of leaching of bases from the soil. The activities of some organisms may have an indirect effect on pedogenesis. For example, in the arctic tundra during the long winters lemmings (*Lemmus sibiricus*) live in the grass and sedge layer between the frozen soil and the snow cover. During the brief summer the snow melts and the permafrost thaws allowing plant growth. Grazing of the vegetation cover by the lemmings removes this insulating layer from the soil surface allowing the soil to thaw to a greater depth than if the lemmings were absent, thereby increasing biological and chemical activity in the soil.

4.5.1 Role of microorganisms

Microorganisms are responsible for some features characteristic of particular soil profiles. Under anaerobic conditions the reduction of Fe^{3+} (red and insoluble) to Fe^{2+} (dull grey/green and soluble) during periods of waterlogging produces mottles characteristic of gley soils. It is not clear whether this effect by microorganisms is indirect via a reduction in O_2 concentration or direct due to the use of Fe^{3+} as an alternative electron acceptor under anaerobic conditions (section 3.2.1). Under acid conditions *Thiobacillus ferrooxidans* produces basic ferric sulphates and weathers minerals that supply K^+ ions to produce the mineral jarosite. These processes contribute to the formation of acid sulphate soils.

Microorganisms participate in the development of soil structure. The stabilisation of soil aggregates, which are formed largely as a result of physical forces, is important for soil development and for crop growth, as it determines the water and air relationships for that soil (see chapter 1). In most soils, organic binding agents are involved; they may be decomposition products of plants, animals or microorganisms, microbial cells or products of microbial metabolism.

The addition of organic matter to soil in the presence of microorganisms improves soil physical conditions. The more readily available organic matter produces a more rapid response. However, there is no simple relationship between microbial numbers and the extent of soil stabilisation. Polysaccharides appear to be the major factor in aggregate stabilisation, particularly in cultivated soils. Bacteria can produce polysaccharides in laboratory culture and it seems likely that they also do this in soil, perhaps using root-derived material as substrates. Bacterial polysaccharides sometimes appear resistant to decomposition in soil. This may be due to the presence of metal cations, or to protection from microbial attack if the polymers are within aggregates.

Fungal polysaccharides are of lower molecular weight than bacterial polysaccharides, and a decrease in the molecular weight of the polymer is often associated with a decrease in adsorption energy. Polysaccharides extracted and fractionated from soils comprise a complex mixture of sugar units. The binding activity of polysaccharides is due to the length and linear structure of the molecule which allows it to bridge spaces between soil particles. Their flexibility allows many points of contact so that van der Waals forces can be more effective. Also, the hydroxyl groups allow hydrogen bond formation, and the acid groups allow ionic binding to soil particles via multivalent ions (cation bridges).

Bacteria are readily adsorbed to the surface of soil particles. The chemical functional groups present on the bacterial cell surface probably allow a mechanism of binding similar to those for organic polymers described above.

Bacteria surrounded by clay crystals could form a microaggregate whereas fungi are probably responsible for binding larger soil particles. Fungal hyphae retain their strength after they die allowing a more permanent effect. Fungi may also form aggregates by bringing soil particles together as they grow. Figure 4.8 represents the possible effects of microorganisms on aggregate stability.

While smaller aggregates (2–20 μm) are probably bound together by organic bonds, larger aggregates (>2000 μm) may be held together by a network of roots and fungal hyphae. Plant roots increase the stability of the

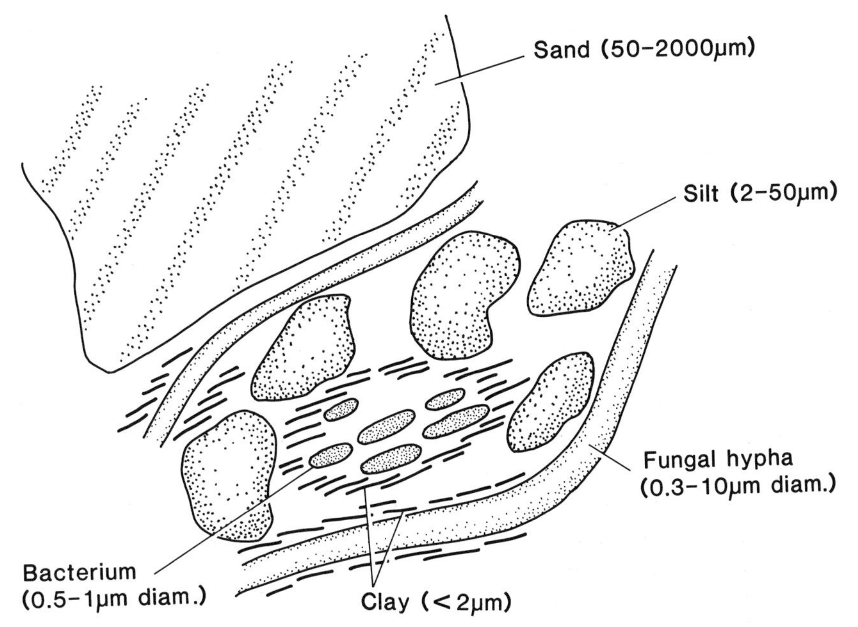

Figure 4.8 Schematic diagram of microorganisms binding soil particles. After J.M. Lynch and E. Bragg (1985).

surrounding aggregates and this effect may be partly due to mycorrhizal hyphae. It has been observed that lucerne (*Medicago sativa*) when grown as a break crop improves the structure of soil.

4.5.2 *Role of mesofauna and macrofauna*

The burrowing activity of some soil animals has important consequences for pedogenesis. The term crotovina was introduced by Russian workers to describe the burrows found in the black chernozem soils of the Russian grain belts. These are formed by the inwash of surface soil into the burrows of hibernating animals such as marmots.

A cicada crotovina, a cicada burrow filled with material from the horizon in which the burrow occurs, has been described for certain soils in the western USA where there is extensive burrowing by these animals. Newly hatched nymphs of species such as *Platypedia* sp. drop to the ground, burrow into the soil, attach themselves to plant roots and feed on plant sap. Their life cycle lasts for 2–6 years. The nymphs are most numerous at depths of between 30 and 100 cm, with the smaller ones deeper in the soil. The nymphs inhabit the open portion of the burrow, up to 75 mm long and 20 mm in diameter. They move through the soil to find food, to escape from low temperatures and to emerge from the soil when mature. They continually backfill the burrows with local material, and up to 75 cm of the profile may be composed mainly of cicada crotovinas. Calcium carbonate may accumulate with time leading to the formation of cemented nodules and cylinders. A cylindrical blocky soil structure is produced as a result of this activity. Cicada crotovinas are less common in soils with high bulk density and a textural B horizon. They prefer well-drained silt loams, and are often associated with shrubby plant species.

A major role for termites and earthworms in profile development has been proposed for a sequence of soils near Ibadan in Nigeria. A sequence of profiles taken across a stream valley developed from granite gneiss and supporting rain forest demonstrate the inter-relationship between soils developing on a slope (Figure 4.9). Such a sequence of soils is termed a catena. Above the sedentary horizon (S) lies a horizon of soil creep (Cr). This creep horizon is subdivided into an upper horizon (CrW) comprising humus and material derived from worm casts, a middle horizon (CrT) formed by the action of termites, and a lower horizon (CrG). This CrG horizon comprises quartz gravel concentrated by eluviation of clay into the S horizon and the removal of fine earth fragments into the horizon above by termites. The largest size of particle found in the CrT horizon is 4 mm, which corresponds to the largest size particle that the termites can carry or digest. Worms sort the fine material (maximum size 0.5 mm) from the CrT horizon and deposit it on the soil surface. The worm cast material and topsoil derived from it is estimated to be 30 kg m^{-2}. Where worms are absent from the soil the topmost layers are formed largely of earth deposited by ants.

The location of the profiles on a slope means that they are in a continuous state of development in which the upper horizons of the soil's up-slope are being removed by creep and constantly being renewed by material from the top of the sedentary horizon due to the activity of termites and worms. Soils lower down the slope accumulate material (Figure 4.9). The termites are particularly important here in the formation of a gravel-free topsoil. Also, their activity disturbs the topsoil and accelerates creep causing a continual lowering of the topsoil, allowing the tree roots to maintain close contact with the nutrients released in the subsoil.

4.5.3 Role of plant roots

Roots are important sources of substrates for soil organisms, and the rhizosphere provides a localised environment in which microbial activity is increased. However, soil physical and chemical conditions in the rhizosphere may also be quite different from conditions in non-rhizosphere soil. In addition to the indirect effects of roots on soil properties, for example O_2 consumption and polysaccharide production by microorganisms utilising root-derived material, roots may directly affect the rhizosphere soil.

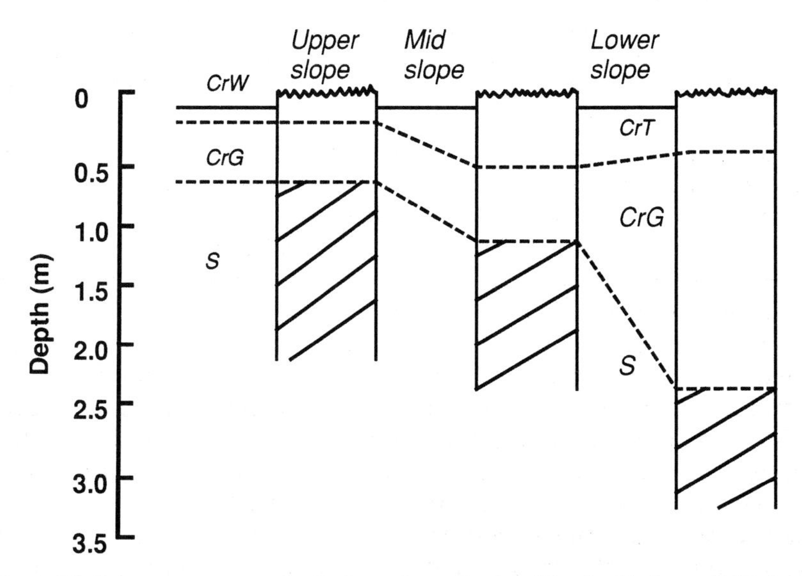

Figure 4.9 Catenary sequence of soils formed near Ibadan, Nigeria. Horizons: CrW, formed by worms; CrT, formed by termites; CrG, gravel accumulation; S, sedentary horizon. After P.H. Nye (1955) *Journal of Soil Science* **5**, 7–21.

Roots generally follow pores and channels that are not much less in diameter than their own; in smaller channels they do not grow freely unless some soil is displaced as the root advances. For example, a pea (*Pisum sativum*) root penetrating clay compacts the soil and orients clay particles within 1 mm of the root surface. In larger channels, such as those formed by earthworms, poor hydraulic contact may exist between the root and soil, restricting the uptake of water. This effect of lack of contact may be reduced by the proliferation of root hairs in the humid air of large soil pores. Even though the pores of diameter greater than 10 μm are drained at −30 kPa a thin film of water is always present on the root and root hair surfaces. Such a film of water could be important in allowing movement and growth of microorganisms in otherwise dry soil.

The hydrophilic mucilages secreted by roots, which are attractive to water, form a layer 1–5 μm thick which assists in the formation of a bridge across which ions and neutral solutes may pass. This could allow plants to take up water which would otherwise be unavailable in dry soils. Mycorrhizal hyphae may also assist in maintaining contact between roots and soil. Such mechanisms will be particularly important as the soil dries out, for instance wheat roots can shrink by 60% of their original diameter at −1000 kPa. Surface-active materials may be produced by roots or the rhizosphere microflora which can change the surface tension and hence the water potential at a particular water content.

Simulation models predict a lower water content around the root compared to the bulk soil, particularly in a dry soil (hence a low hydraulic conductivity) with a high demand for water by the root. However, such predictions depend greatly upon the water uptake per unit root length and this is extremely difficult to measure. The most rapid uptake of water occurs 10–100 mm behind the root tip in solution culture. The interfacial resistance (and hence gradient in water potential) between root and soil may not be important until the soil moisture content is less than 20% of its saturation value.

There may be situations in dry soils in which the water potential of the rhizosphere soil is lower than that of the root, and water therefore moves from the root into the soil. This could happen with a deep-rooting plant in a dry soil at night time when stomata are closed, and the water potential of the soil around roots in the surface layers is lower than the water potential of the soil around deeper roots. In such circumstances, the roots would provide a relatively low resistance pathway for water movement from wet to dry soil. There is no evidence that the resistance to flow of water from roots is different from the resistance of flow of water into roots.

A similar mechanism has been proposed for the desert tree *Prosopis tamarugo* which thrives in the deserts of northern Chile in which the groundwater is often over 40 m deep and no contact occurs with the root system. When the relative humidity is high the plant absorbs water through its leaves and this water then flows down to the root system and into the rhizosphere where it is stored and re-absorbed as required. Such changes in soil water content around roots could have important consequences for the uptake of nutrients such as phosphate, for local growth and activity of roots, and for the activity of the rhizosphere microflora in dry soils.

It is essential that plants maintain electrical neutrality across the root surface in response to the uptake of cations (positively charged) and anions (negatively charged). Analysis of plants indicates that if N is absorbed as NO_3^- then plants generally absorb more anions than cations. This should lead to an outflow of anions, commonly HCO_3^-. If plants absorb N as NH_4^+ or as N_2 gas (N_2 fixation) then the cation:anion balance indicates that cations, usually H^+, will move out. This is confirmed in experiments in which plants utilising NO_3^- cause an increase in the alkalinity of the rhizosphere soil, whereas plants utilising NH_4^+ or N_2 cause localised acidification. Differences of up to 1 pH unit have been reported between rhizosphere and non-rhizosphere soil. The pH change at the root surface will depend upon the rate of release of HCO_3^- and H^+, the soil pH buffer capacity and other factors.

The rhizosphere of rape (*Brassica napus*) has been studied in detail because of the plant's ability to take up phosphate from phosphate-deficient soils. This appears to be due to the acidification of the rhizosphere (by up to 2.5 pH units) and increased solubilisation of P by the plant in response to reduced uptake of NO_3^- under conditions of low phosphate. Different nutrients may be taken up by different parts of the root; this may cause the rhizosphere pH value to vary along the root.

In addition to the outflow of bicarbonate ions, CO_2 is liberated at the root surface by respiration. This is unlikely to cause a local acidifying effect because under aerobic conditions CO_2 diffuses rapidly away from the root through the air-filled pore space. In contrast, the bicarbonate ion is confined to the soil solution where its mobility is 10^4 times lower than gaseous CO_2. In calcareous soils the effect of CO_2 may be significant.

The concentration of other ions in the rhizosphere may be affected by plant uptake and supply from the surrounding soil. The major transport processes occurring near the root surface are shown in Figure 4.10. Rapid equilibrium takes place between solutes in the soil solution and those absorbed on the adjacent solid surfaces. These adsorbed solutes tend to buffer the soil solution against changes in concentrations induced by root uptake. It is not clear whether the rate of absorption of solutes by roots is determined by their concentration in soil solution or by their concentration within the plant.

Solutes move to the root by mass flow (movement of the soil solution caused by transpiration) and by diffusion (movement of solutes in response to a concentration gradient), with both processes operating simultaneously. Calculations indicate that more than sufficient Ca and Mg (except in sodic

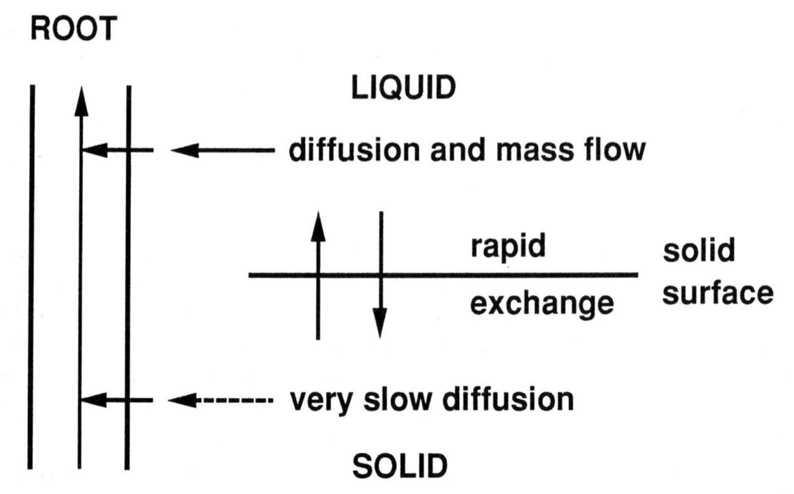

Figure 4.10 Solute transport processes near an absorbing root. After P.H. Nye and P.B. Tinker (1977).

or very acid soils) and Na to satisfy the plant demand is transported to the root surface by mass flow, but insufficient K, and especially P.

If a solute is absorbed at a fast rate relative to water, for example P and K, then the solution concentration at the root surface decreases. Some of this concentration change may be buffered by the release of ions from solid surfaces, but the remaining concentration gradient will lead to diffusion of ions towards the root. If water is absorbed at a fast rate relative to the solute then the solute concentration at the root surface increases, leading to a localised increase in salinity and a decrease in osmotic potential.

These effects may be reduced by the diffusion of solutes away from the root. Therefore, the processes of mass flow and diffusion can lead to zones of nutrient depletion or accumulation around the root. The low concentration of phosphate in the rhizophere (<1 μM) may have important consequences for colonisation of the root region and microorganisms with reserves of polyphosphate may be favoured.

4.5.4 Role of vegetation

The complex interactions that occur between soil-forming factors during pedogenesis make it difficult to clearly identify the major causes of profile change. In some cases the dominant factors are more obvious; in the freely drained soils of north west Europe, climatic factors cause progressive leaching and consequently increased acidity leading in some soils to podzolisation. However, it has been known for a long time that changes in vegetation are also important. For example, if oak (*Quercus* sp.) is replaced by beech (*Fagus sylvatica*) or heather (*Calluna vulgaris*) the brown forest soil with mull humus (neutral pH, high base content and earthworm activity) changes to a podzol soil with mor humus (acid, with a mainly fungal population, and with no mixing of organic and mineral soil).

Different plant species and their litter vary in their content of nutrients and secondary chemicals, which in turn can affect the rate of decomposition and may directly affect soil acidity, and accelerate eluviation (movement of soil material through the profile). Species also differ in the way in which they modify the composition of rainfall draining off the leaves. Alder (*Alnus* sp.) tolerates acid soils and also acidifies rapidly, with decreases in soil pH values of 0.5–2.0 units recorded in 20–30 years. Similar changes occur under heather (*Calluna vulgaris*) and gorse (*Ulex* sp.). It may take 1000 years for a mature podzol to form, but the early stages of horizon differentiation occur within less than 100 years.

Some species may reduce soil acidity and reverse the trend towards podzolisation. The colonisation of heathland by birch (*Betula* sp.) may increase the soil pH value from 3.8 to 4.9 in 90 years. Bracken (*Pteridium aquilinum*) sometimes reduces acidity and at other times has the opposite effect. Herbaceous species generally do not acidify soils. For example, heavily grazed natural pastures of *Agrostis* sp. and *Festuca* sp. are associated with mull-like humus, partly due to the rapid cycling of litter and nutrients by the grazing animal which prevents organic matter accumulation.

A marked increase in soil acidity and consequent decline in productivity of subterranean clover (*Trifolium subterraneum*), an annual self-regenerating pasture legume, has been observed for a range of soils and farming systems in south-east Australia. Decreases of 1 pH unit have been recorded over 25–50 years. This is due to a combination of factors, including the accumulation of organic matter giving rise to an increase in cation exchange capacity and exchangeable acidity, and acidification of the legume rhizosphere.

The effect of vegetation on soils is evident at a site in Saskatchewan where, on uniform parent material with only small differences in climate, grassland is found adjacent to woodland on gently sloping hills. The chernozem soil formed under grassland was changed to a podzol following invasion by trees which caused a decrease in pH, an increase in leaching of $CaCO_3$, the appearance of litter and humus horizons, and an increase in clay migration through the profile.

A succession of plant species is often observed in natural ecosystems; species-poor associations of low stature such as grasses and sedges are followed by communities of tall plants with many species and complex structures. A link is often assumed between plant succession and pedogenesis, with a natural progression toward a plant climax community on a mature soil. If such a climax community does occur, then if it is undisturbed it may be assumed to be in permanent equilibrium with the environment, to have maximum total biomass, highest species diversity and highest stability (see also chapter 6). However, if the environment does not remain constant then such a situation will not occur.

A study of five dated river terraces in the Alaskan tundra, all with comparable parent material, climate and topography but formed at different times, provides an example of changes in vegetation and soil with time. The pioneer vegetation (25–30 years old) comprises isolated low shrubs such as *Dryas* sp. and herbaceous plants such as *Astragalus* sp., both of which are N_2-fixing plants. On the next older terrace (100 years old) the pioneer shrubs were replaced by willow (*Salix* sp.) and a mat of plants including grasses such as *Poa* sp. A 5 cm deep organic layer of decomposing moss and plant tissue has formed. During the shrub stage (150–300 years old) willow was replaced by birch (*Betula* sp.) and other trees which survive by putting adventitious roots into the moss layer, which is now up to 35 cm deep. The tundra vegetation on the oldest terrace (5000–9000 years old) comprises low shrub-sedge tussock moss and cotton grass (*Eriophorum vaginatum*). The plant succession is associated with an increase in the soil N content.

The succession of plant species on sand dunes is often paralleled by a microbial succession in the soil. On the shores of Lake Michigan (USA) no characteristic microflora was associated with the pioneer species, *Ammophila breviligulata* (marram grass), possibly due to the instability of the sand and the low organic matter content. However, a distinct fungal flora was associated with the later community of *Andropogon scoparius* (bunchgrass).

A succession on sand dunes in Aberdeenshire (Scotland) from *Ammophila arenaria* to *Calluna vulgaris* (heather) was associated with an initial increase in the numbers of bacteria and fungi followed by a decrease in the numbers of bacteria while the fungal population continued to increase (Table 4.5). The reduction in the bacterial population was probably due to an increase in the soil acidity during the later stages of succession. In the early stages of succession, when the soil organic matter content is low, the input of organic material from roots and the rhizosphere microflora is particularly important.

4.6 The influence of man

Man can have a major effect on soil development and the effect is often rapid. For example a marsh can be converted into a meadow and a tropical rain forest into an unproductive pasture within a few years. The effect of man is most often via his effect on vegetation, usually associated with the introduction or modification of agricultural practices.

Table 4.5 Changes in microbial populations and acidity in soil associated with a succession of plant species on sand dunes in Scotland. After D.M. Webley *et al.* (1952) *Journal of Ecology* **40**, 168–178

Vegetation	pH	Bacteria (number per gram)	Fungi (number per gram)
Open sand	6.8	18 000	270
Yellow dune	6.7	1 630 000	1 700
Early fixed dune	5.1	1 700 000	68 470
Dune pasture	4.8	2 230 000	209 780
Dune heath	4.3	127 000	148 000

The link between civilisation and soils is illustrated by the rise and fall over a period of 4000 years of the Mesopotamian, Assyrian and Babylonian civilisations in the fertile crescent of the Tigris and Euphrates river basins in present-day Iraq. The lower reaches of these two rivers were colonised around 5000 BC by the Sumerians. This alluvial plain, dusty when dry and swampy when wet, was transformed by dykes and canals into extensive irrigated fields and date-palm plantations. Civilisation flourished: writing was invented, together with the wheeled vehicles, the animal-drawn plough and devices for irrigation. However, around 3000 BC the civilisation went into decline and the centre of power moved northwards into Babylonia and Assyria.

The causes of this decline in the Sumerian civilisation are not clear. It was partly due to the accumulation of silt carried down from the Armenian mountains and deposited in the fields and blocking the irrigation system. The practice of flood irrigation using the waters of the Euphrates caused the water table to rise; salinity increased in the soil as a consequence. Wheat was the preferred grain, but there is evidence of an increase in the use of barley which is more tolerant of salt.

In contrast, the river Nile has supported over 5000 years of civilisation without interruption. The seasonal flow of the Nile and its load of silt and humus carried from the sources of the Blue Nile and White Nile have provided conditions more suitable for sustained food production than those in Mesopotamia. The irrigated farming systems in the narrow flood plain along the length of the Nile allow up to four crops to be produced per year, and can support a stable population of about 1.5–2.5 million; the population of Egypt today is 55 million.

Moving away from irrigated land, man can have a major influence on soil organic matter content, usually as a consequence of changing the vegetation cover. Management practices can alter the annual input of organic matter, for example when crop residues are removed from the soil or when vegetation is burned, and can also alter the rate of turnover of organic matter, for example by the cultivation of soil.

Table 4.6 Changes in soil properties under different soil and land management practices at IITA in Nigeria and in Vindhyan hill tract Uttar Pradesh, India. After A. Ayanaba *et al.* (1976) *Soil Biology and Biochemistry* **8**, 519–525; S.C. Srivastava and J.S. Singh (1991) *Soil Biology and Biochemistry* **23**, 117–124

Country	Management	pH	Total C ($\mu g\ g^{-1}$)	Total N ($\mu g\ g^{-1}$)	Biomass C ($\mu g\ g^{-1}$)
Nigeria	Bush regrowth	7.0	15 200	1320	340
Nigeria	Maize/+ res[a]	6.2	14 300	1180	200
Nigeria	Maize/− res[a]	5.5	11 100	920	90
India	Forest	6.4	21 800	2236	609
India	Savannah	7.0	12 050	1065	397
India	Cropped land	6.7	10 650	1056	250

[a]res = above-ground residues
NPK fertiliser applied to both maize treatments

In Ghana, when an evergreen forest was cleared and cropped with a maize (*Zea mays*)/cassava (*Manihot esculenta*) rotation for 8 years, the C content of the soil was reduced from 2.2% to 1.5%. Similar data have been reported for Nigeria and India (Table 4.6). The addition of crop residues to the soil in Nigeria increased the total C, total N and microbial biomass C compared to the soil which received no residues. Conversion of dense dry tropical deciduous forest into savannah or cropland on an Ultisol in northern India caused a halving of the organic C and N, and of the biomass C and N.

In these examples, the amount of soil microbial biomass changes in proportion to the changes in amount of soil organic matter. However, data from Denmark for a sandy loam which had received either barley straw residues or no crop residues for 18 years indicated that the crop residues increased organic C by 5% but the biomass C by 37–45%. From these data it was concluded that changes in the amount of soil microbial biomass may provide an early indicator of changes in soil organic matter content.

The cultivation of a virgin prairie soil in Kansas for 40 years reduced the amount of N in the soil by almost 50%. The introduction of arable farming nearly always leads to a reduction in the soil organic matter content. Conversely, the organic matter content of an arable field at Rothamsted Experimental Station left uncultivated (fallow) for 90 years increased almost three-fold as woodland regenerated.

Modern temperate agricultural methods may be able to sustain production at low soil organic matter levels but this depends upon an input of organic matter from crop residues. In the tropics, however, a fallow period is essential to allow organic matter and nutrient contents of the soil to recover before further cropping.

The natural climax vegetation in the humid temperate zone is broad-leaved woodland associated with a brown earth soil. Evidence obtained from pollen grains preserved in acid soils, and from estimates of mean residence time using ^{14}C-dating of podzol soils in eastern England, indicates a change in soil development caused by the clearance of woodland by Neolithic and Bronze Age man. The heavy canopy of a broad-leaved forest transpires water taken from deep in the soil and also reduces soil evaporation. The removal of this vegetation causes increased drying of the soil surface and a rise in the water table. In Western Australia this is causing an increase in salinity of soils cleared of the natural forest. Exposure of the soil surface to the atmosphere also causes an increase in the rate of organic matter decomposition. The absence of an input of leaf litter with high base content causes a reduction in the base content of the topsoil and together with the reduced organic matter content this reduces the earthworm population. Podzolisation normally occurs as a consequence, but this may be prevented by cultivation, which brings bases to the soil surface, and by the addition of organic manures and fertiliser.

It is clear therefore that man is now one of the major factors in soil formation and development. In the most extreme case man's activities can lead to the loss of soil as in the most serious cases of soil erosion. This topic and the related one of desertification are considered further in chapter 7.

5 Environmental issues

5.1 Introduction

Environmental considerations are playing an increasingly important role in determining management strategies for soil and land. In the so-called developed countries issues such as nitrate leaching are now of greater prominence in policy development than are the more traditional issues such as food production.

In many cases, the concern is not that man has initiated a new process, but that the rate of a naturally occurring process has been accelerated and that the normal cycle of processes has been disturbed. In some cases, for example soil erosion, the process can be seen to be leading to a downward spiral, with potentially disastrous consequences.

Many important environmental issues involve aspects of the biology of soil. These issues cannot be satisfactorily considered in isolation from a general understanding of the subject as a whole. Therefore the information included on selected environmental issues in this chapter builds upon the material presented in the earlier chapters. Shortcuts should be avoided as they may ultimately lead to delays in understanding.

5.2 Acidification

Acid soils occur widely around the world, particularly in humid tropical regions, for example the cerrados of Brazil where soils have been strongly leached over a long period of time. Acidification is a natural process, and in the absence of basic parent material, is the normal course of development for soils.

The processes of weathering cause soils to become more acidic, and the intensity of the acidity, as measured by soil pH, is affected by the type of minerals in the parent rock, climate and vegetation. Acidity is generated by CO_2 dissolving in the soil solution, and by nitrification and oxidation of S. Where fertilisers such as $(NH_4)_2SO_4$ are used, acidity is generated by nitrification of NH_4^+, and bases such as Ca^{2+} and Mg^{2+} are lost from the soil during leaching of the NO_3^- produced. Plant roots, plant litter, and carboxyl and phenolic groups on humified organic matter also contribute to soil acidity.

The increase in use of fossil fuels in developed countries has led to an

Table 5.1 Estimated inputs of acidity into UK soils. After D.L. Rowell and A. Wild (1985) *Soil Use and Management* **1**, 32–33

Source	Annual input of H^+ ($g\ m^{-2}$)
Natural	
CO_2 in soil pH > 6.5	0.72–1.28
Organic acids in soils and from vegetation	0.01–0.07
Acid rain	
Wet deposition	0.03–>0.10
Dry deposition	0.03–>0.24
NH_3 and NH_4^+ oxidation	0.07
Land use	
Cation excess in vegetation	0.05–0.20
NH_4^+ oxidation and NO_3^- leaching (fertiliser)	0.40–0.60
Oxidation of N and S (from organic matter and leaching)	0–1.00

additional input of acidity into soils from atmospheric pollutants, mainly SO_2 and various oxides of N and NH_3, often referred to as acid rain. Estimated inputs of acidity into UK soils are shown in Table 5.1.

As soils become more acidic, basic cations are displaced from exchange sites and leached down the soil profile, and exchangeable H^+ ions take their place on the clay minerals and organic matter. Clays with appreciable amounts of exchangeable H^+ are unstable however, and slowly dissolve, releasing Al, Mg and silica; Al becomes the predominant exchangeable cation. Eventually, after prolonged weathering the clay minerals are destroyed and mainly gibbsite, silicates and various iron oxide minerals remain. Soil acidity is therefore different to the acidity of a solution because it is controlled mainly by ion exchange reactions involving both inorganic and organic soil components.

A range of chemical factors is associated with soil acidity. Acid soils contain large amounts of exchangeable Al and small amounts of exchangeable bases, particularly Ca ions. The solubility of Mn also increases under acid conditions. The equilibrium with the soluble divalent form of Mn is controlled by the redox potential. In very acid soils which are waterlogged high amounts of exchangeable Mn^{2+} and Fe^{2+} may occur. The amount of readily available phosphate may be low in acid soils as phosphate is adsorbed onto the surface of positively-charged Fe and Al oxides and hydroxides. These acidity factors may have a greater effect on biological activity in soil than does the concentration of H^+ ions. The adverse effects of soil acidity can be overcome by the application of lime to soil.

5.2.1 Aluminium toxicity

Acid soils often contain appreciable amounts of Al, due to the ubiquitous nature of this element in soil (it is the third most common element in the earth's crust), and to the increase in solubility which occurs as acidity increases. Al has been shown to be the most important factor related to acidity that affects growth of many plants including lucerne (*Medicago sativa*) in Australia, cotton (*Gossypium* sp.) in the southeastern United States and coffee (*Coffea* sp.) in Brazil.

Surface soils often have lower amounts of exchangeable Al than subsoils at pH values less than 5, due to the greater affinity of Al for adsorption sites on organic matter than on clay minerals, and additions of organic matter to acid soils can often allow plants to grow satisfactorily. The problem of subsoil acidity is a particularly serious one; some success has been found in the use of gypsum (likely to be in greater supply as a by-product of desulphurisation processes incorporated into coal-fired power stations).

The symptoms of Al toxicity are similar in most plants. The roots are stunted and thickened with the lateral roots appearing as peg-like projections. Aluminium may interfere with P uptake and metabolism within the plant root, it may interfere with mitosis, and it may displace Ca from the root cell walls preventing cell growth. Some plant species such as tea (*Camellia sinensis*) are able to tolerate high concentrations within the plant by using chelating agents which allow safe transfer of Al to the older leaves where it may accumulate in concentrations up to 20 000 $\mu g\ g^{-1}$.

Little is known of the mechanisms of Al toxicity to soil microorganisms despite the fact that this may be the major factor limiting microbial growth and activity in acid soils. The microbial component of the legume–*Rhizobium* symbiosis is more sensitive to acidity and Al than the host plant, and Al appears to enter the bacterial cell and interfere with DNA replication. Species of *Rhizobium* vary in their sensitivity to Al, but the mechanisms giving rise to tolerance of Al are not known.

5.2.2 *Protected microsites*

Bacteria isolated from acid soils when tested under laboratory conditions often do not appear to be tolerant of pH values corresponding to the pH of the soils from which they were isolated. This has been observed for *Streptomyces* sp., *Arthrobacter* sp., *Rhizobium leguminosarum* biovar *trifolii* and *Nitrobacter* sp.

An acidophilic population (one that prefers acid conditions) may exist in an acid soil, but the procedures used to isolate the organisms from soil may favour neutrophilic organisms (those that prefer neutral conditions). Alternatively, there may be protected microsites in the soil which have a pH value higher than that of the bulk soil. The acidity of the rhizosphere soil can be quite different from that of the non-rhizosphere soil and localised increases in acidity may occur at the surface of clay minerals. Several distinct population may co-exist in the same soil. For example,

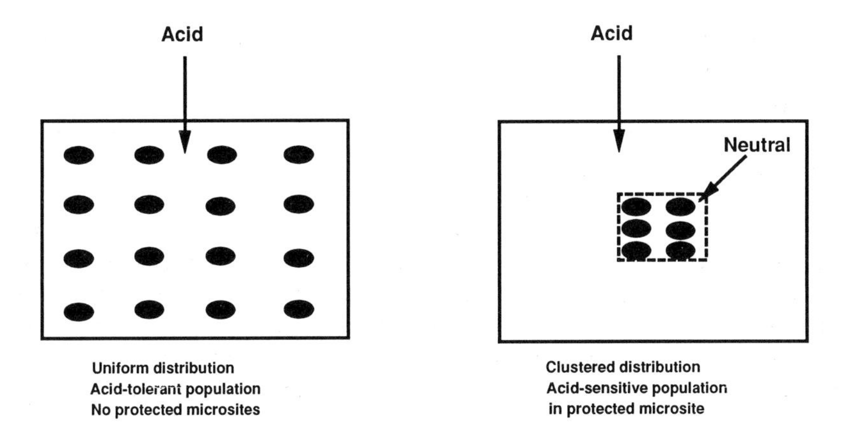

Figure 5.1 Hypothetical scheme for the survival of acid-tolerant and acid-sensitive bacteria in an acid soil.

only neutrophilic species of *Nitrosomonas* were isolated from an acid forest soil in the USA, but both acidophilic and neutrophilic species of *Nitrobacter* were isolated from the same soil.

Figure 5.1 illustrates the possible role of microsites in the ecology of bacteria in soil. With such a highly heterogeneous medium as soil, pH measurements only indicate average values; it is not possible at present to measure localised pH values in soil around plant roots or particles of decomposing organic matter to test the hypothesis of protected microsites.

Bacteria may be able to modify their local environment to produce favourable microsites. For example, it has been proposed that root nodule bacteria present in acid tropical soils are adapted to those soil conditions because they are often slow-growing organisms (*Bradyrhizobium* sp.) which produce alkali in laboratory media. However, it has not yet been demonstrated that these organisms produce alkali in soil, nor that the amount of alkali produced would be large enough to bring about a change in pH of the soil surrounding a bacterial cell or colony of cells. Fungi are generally more tolerant of soil acidity than bacteria, but the reasons for this are not clear.

5.2.3 Acid rain

There is concern over the effects of atmospheric pollutants, in particular ozone and oxides of S and N, on plants and soils. This was stimulated by observations of widespread damage to forests in central Europe and in the northeastern USA. A wide range of species is affected over a large area, and the typical symptoms in Norway spruce (*Picea albies*) are needle discoloration and crown thinning. Increasing damage has been observed in broad-leaved species such as beech (*Fagus sylvatica*) and oak (*Quercus* sp.). There are regional differences in forest damage, and if trees are not severely affected the damage is reversible.

A range of damage types is now recognised. However, the simultaneous appearance of these different types of damage within a short space of time over almost the whole of the former West Germany indicates a common, as yet unknown, factor. There are several current hypotheses concerning the causes of this damage. The first involves multiple stress in which air pollution impairs plant metabolism and makes the plant more susceptible to other stresses such as climatic factors or nutrient deficiency. This hypothesis is difficult to test. Soil acidity may be a contributory cause in which Al impairs the tree root system. However, damage has been observed on trees growing on a range of soils including calcareous soils in the Alps. Ozone may interact with acid mist to induce or exacerbate nutrient deficiencies.

Magnesium has been implicated in the high-altitude spruce damage; Mg concentrations in these soils are marginal for tree growth and a series of dry years may have reduced the Mg supply to the trees below a critical level. Finally, in The Netherlands excess annual ammonia deposition (NH_3–N) in the order of 5 g m^{-2}, derived mainly from animal slurries, has contributed to an increase in the susceptibility of pines to climatic and biotic stresses, mainly due to the acidity produced from NH_4^+ when it enters the soil. It is unlikely, therefore, that a single factor is responsible for the range of types of forest damage observed.

The estimates for inputs of acidity into UK soils (Table 5.1) show that wet and dry deposition together may account for as much acidification as that produced by vegetation and nitrification.

In view of the information presented it is difficult to draw any general conclusions regarding the effects of acidification on soil biological processes. It seems fairly clear that earthworm populations are dramatically reduced in acid soils, however, they may not be completely absent.

The microbiology of acid soils is particularly difficult to understand. Many studies have examined the short-term effects of simulated acid rain on biological processes in soil, but the results of these studies may not be particularly relevant to long-term effects on ecosystems. Microorganisms are capable of adapting to a certain degree of increased acidity in soils, and the possibility of protected microsites complicates the interpretation of experiments. The autotrophic bacteria are particularly intriguing from this point of view: some S oxidisers such as *Thiobacillus ferrooxidans* require acid conditions in order to survive, yet paradoxically the ammonium oxidisers (which generate acidity) and the nitrite oxidisers appear particularly sensitive to acidity.

5.3 Salinity

Salinity has been a threat to soil and therefore to civilisation since the very first settled agriculture.

Soluble salts (mainly Na_2SO_4 and chlorides of Na and Ca) accumulate in the surface of soils under hot dry conditions when the groundwater comes within a few metres of the soil surface (as happens after forest clearance) to form saline soils. During dry periods, when evapotranspiration is less than rainfall, the surface of saline soils is covered with a salt crust which dissolves in the soil solution each time it is wetted. Saline soils may also be produced by poor irrigation practice. Irrigation water always contains some dissolved salts, and these salts remain in the soil after the water has evaporated or been transpired by the plant. It is important therefore to ensure that salts are leached from the root zone using water in excess of that required by the crop, and to ensure that drainage is provided.

If the salts are mainly Na, then as they are washed out, $NaHCO_3$ is formed which causes the soil pH value to increase above 9. In these sodic soils the Na and high pH cause humic colloids to deflocculate and clays to swell or disperse, as the normal attractive forces become forces of repulsion, leading to an unstable soil structure.

Salinity causes dwarfed and stunted plants; this may be due to a decrease in water availability because of the decreased water potential (increase in osmotic pressure) caused by the high salt concentration. The effect of salts in decreasing the water potential may cause symptoms of drought. Some plants adapted to saline soils are better able to extract water from dry soils, but this is not always the case. For example, the coconut palm (*Cocos nucifera*) is salt tolerant but not drought tolerant.

In addition to this osmotic effect, salts may also exert a specific ion effect. Many fruit trees are sensitive to high concentrations of Na and Cl ions, which may interfere with metabolism and nutrient uptake. However, in general the effects of salts on plants depend upon the total concentration of soluble salts rather than on specific ions.

Plants tolerant of salinity must be capable of either excluding salts or adjusting to osmotic stress. The latter is achieved by the uptake and accumulation of inorganic ions such as potassium, or the synthesis of low molecular weight organic acids such as malate. Plants are generally more sensitive during the seedling stage. Date, cotton (*Gossypium* sp.) and barley (*Hordeum vulgare*) are examples of salt tolerant plants, whereas beans (*Phaseolus vulgaris*) and white clover (*Trifolium repens*) are sensitive.

Soil microorganisms face similar problems to plants in saline soils. Although many biochemical functions require particular inorganic ions, an increase in concentration above normal intracellular concentrations may lead to disruption of cell function by reducing enzyme activities. In order to prevent osmotic dehydration and to maintain optimum turgor pressure, most bacteria under osmotic stress accumulate K ions and low molecular weight organic compounds such as amino acids, betaine, trehalose and glycerol.

Gram-negative bacteria accumulate glutamate using glutamate dehydrogenase (which is sensitive to increases in pH value), and electrical neutrality is maintained by also accumulating K ions. Gram-positive bacteria tend to accumulate proline which does not require K accumulation. The cytoplasmic ionic strength appears to act as a signal which controls the induction of systems for synthesis and accumulation of compatible solutes.

The osmotic effect of most organic compounds such as glycerol is lower than that of NaCl, and complete osmoregulation may not occur in salt-tolerant bacteria. Other adaptations to saline environments may include changes in cell wall phospholipid composition, and the accumulation of betaines which may stabilise the molecular conformation of various enzymes. Clay minerals may also partially protect microorganisms against osmotic stress.

5.4 Heavy metals

Metals have been associated with human activity for thousands of years; copper and lead have been used since prehistoric times, but cadmium was only discovered in the early nineteenth century. Heavy metals differ from most pesticides (which are considered in section 5.7) in two important respects: first they occur naturally in all soils, waters and living organisms; second they are not dissipated by biological means.

Like pesticides, heavy metals are now considered essential to modern life, and as both population and usage per person increase the problems of dispersal and concentration become more serious. The metals are strongly adsorbed to soil components and accumulate in soil, remaining possibly for thousands of years. The consequences of this long-term accumulation are potentially very serious.

Heavy metals may be introduced into soils from fallout from smelters and the dumping of smelter and mine wastes, urban and traffic sources, and from the application of contaminated sewage sludge (particularly in industrial areas). The extent of this pollution can be considerable, for example, approximately 5×10^5 tons (dry weight basis) of sewage sludge are applied to land in the UK each year and this may contain high concentrations of Cu, Zn, Ni, Cd, Pb and Cr, derived from industrial sources (Table 5.2).

In addition, certain sources of rock phosphate (used in the manufacture

Table 5.2 Concentrations of heavy metals (mg kg^{-1} dry weight) in sewage sludges from different countries. After K.G. Tiller (1989) *Advances in Soil Science* **9**, 113–142

Country		Cd	Zn	Cu	Pb	Ni
UK	range	2–1500	600–20000	200–8000	50–3600	20–5300
	median	20	1500	650	400	100
USA	range	2–1100	72–16400	84–10400	800–2600	12–2800
	median	12	2200	700	480	52
Australia	range	2–285	240–5500	250–2500	55–2000	20–320
	median	26	1900	670	420	60

of phosphate fertilisers) contain up to 90 mg kg^{-1} of Cd, although long-term trials in Australia and the US have shown no significant increase in Cd concentration in the main cereal crops. However, most Cd taken up by plants remains in the roots.

Although many heavy metals serve as micronutrients for microorganisms, all are toxic in high concentrations due to their ability to denature proteins. The effects on soil microorganisms in soil are difficult to quantify due to effects of soil minerals, organic matter and dissolved ions on the solubility and reactivity of the metals. Furthermore, microorganisms may decompose polluting organic materials and contaminated litter to release metals.

Resistance to heavy metals may develop in soil bacterial populations, but this may take several years. There is some evidence that bacteria isolated from heavily polluted soils are more resistant to heavy metals than those isolated from less polluted soils. Metal resistance in bacteria is usually determined by genes carried on plasmids (section 5.8.2).

The effects of heavy metals on soil biological processes may persist for decades after the input of pollutant has ceased. In experiments at Rothamsted, soils were treated with sewage sludge (organic matter plus heavy metals) or farmyard manure (organic matter but no heavy metals) for 20 years. This resulted in a three-fold increase in the amounts of Zn, Cu, Ni and Cd in the soil treated with sludge. Growth of grass was unaffected, but the microbial biomass was reduced by 50%, free-living N_2 fixation was reduced by 90% and clover rhizobia were rendered ineffective even 20 years after applications of sewage sludge had ceased. Critical levels of heavy metals in soils have been established, and are discussed further in chapter 6.

In woodland polluted with heavy metals, plant litter accumulates mainly due to a decrease in the activity of soil invertebrates. Measurements of concentrations of heavy metals in soil or vegetation may not be reliable indicators of the concentrations of metals in primary consumers. Invertebrate animals such as earthworms, slugs and snails and woodlice accumulate heavy metals, and woodlice are particularly useful indicators of

Figure 5.2 The spider *Dysdera crocata* attacking a woodlouse (*Porcellio scaber*). Courtesy of Dr S.P. Hopkin, The University of Reading.

the biological availability of metals in polluted soils. The accumulation of up to 1% Cu and Zn in woodlice is likely to present a major deterrent to predators such as the spider *Dysdera* sp. which feeds almost exclusively on this soil animal (Figure 5.2).

A number of plant species occur naturally on soils containing heavy metals. For example, *Agrostis tenuis* grows on mine workings and is tolerant of Cu, Ni, Pb and Zn, although tolerance of the different metals is not necessarily linked. Leadwort (*Minuartia verna*) is only found on lead spoils. Few plants can exclude metals completely, and some accumulate very high concentrations. For example, *Sebertia acuminata* which grows on Serpetine soils in New Caledonia contains 11.2% Ni (wet weight basis). Ericaceous mycorrhizas may increase the tolerance of plants to heavy metals by reducing metal uptake; for example, *Calluna vulgaris* is able to colonise Cu mine spoil.

5.5 Chernobyl and radioactivity

Radioactivity may enter the environment from a variety of sources. Routine releases occur as a result of the reprocessing of nuclear material as undertaken at Sellafield in the UK and Cap de la Hague in France, or due to plutonium production as carried out at Hanford in the USA. There are also accidental releases; some of the more notable ones include releases from nuclear reactors at Windscale, UK in 1957, from Three Mile Island, USA in 1979 and from Chernobyl in 1986. The explosions in Reactor No. 4 of the nuclear power station of Chernobyl in the Ukraine caused the most serious accidental release of radioactivity to date. A 30 km radius zone around the power plant was evacuated immediately after the release, and an area of 2×10^4 km^2 outside this zone has been designated as contaminated.

The burning of fossil fuels releases pre-existing natural radioactivity which would otherwise remain trapped in the Earth's crust. Of much greater concern are the artificial radionuclides created by neutron-induced nuclear fission in nuclear reactors and nuclear weapon explosions. These artificial radionuclides cover all groups of the periodic table, but the more abundant and toxic nuclides include isotopes of the elements caesium, strontium, ruthenium, cerium, iodine and plutonium.

The behaviour of radionuclides in terrestrial ecosystems is determined by the initial chemical form. The global fallout from nuclear weapons testing comprised mainly water-soluble and exchangeable forms of ^{137}Cs and ^{90}Sr, whereas the deposition of these radionuclides after Chernobyl was often in non-exchangeable forms. In such an accident, rainfall is a major factor in the deposition of fallout, and the major problems after Chernobyl coincided with areas of high rainfall, especially over hilly and mountainous regions of Europe.

The properties of ^{137}Cs (half-life 30 years) are similar to those of K; for example its salts are highly soluble in water, although the metabolism differs. The Cs released into the atmosphere at Chernobyl was associated with particles of a mean equivalent geometric diameter of 0.5–1.0 μm. The properties of ^{90}Sr (half-life 28.8 years) are similar to those of Ca. Radiostrontium is of concern because of its tendency to concentrate in the mineral bone of animals.

As a result of bomb tests in Nevada and in the South Pacific there is much information available on the behaviour of radionuclides in arid soils, and in tropical islands. However, this was of little relevance to the fallout from Chernobyl where effects in Arctic tundra ecosystems (particularly uptake by lichens) and in upland areas were important. The addition of ^{137}Cs to soils as soluble compounds results in about 85% becoming bound, but the remainder is available for plant uptake. The availability depends upon the soil type; in Western Europe the presence of organic matter enhances uptake by pasture plants and consequent contamination of animal tissues. The rapid recycling of radiocaesium in upland ecosystems may be due in part to the activity of the soil microflora, e.g. mycorrhizas.

High-level radioactive wastes and transuranic wastes, mainly in the form of spent fuel from reactors and fuel reprocessing and from plutonium production for nuclear weapons, are disposed of by long-term isolation from the environment. Low-level wastes, which are generated by all activities involving radioactive materials, are disposed of by incineration or burial in shallow trenches up to 6 m deep. Soil may also be treated with γ-radiation for experimental purposes.

Many soil insects are killed by relatively low doses of radioactivity (0.01–0.5 kGy) with the most active animals showing the greatest sensitivity. The sensitivity of cells to radiation generally increases with increased metabolic activity. For example, vegetative cells are more sensitive than spores and cysts. Fungi are the most sensitive group of microorganisms. Viability is decreased by radiation doses as low as 0.025 kGy and *Penicillium* sp. and *Trichoderma* sp. are particularly sensitive. 10 kGy eliminates actinomycetes and 25 kGy sterilises most soils completely. The activity of many soil enzymes shows little decrease after a radiation dose sufficient to inactivate all microorganisms, and this technique is useful in studying extracellular enzyme activity in soils.

Many microorganisms are readily able to colonise soil following irradiation, although nitrification activity is not always re-established. The lysis of cells during and after irradiation leads to an increase in available nutrients such as N, which often results in increased plant growth. There may also be decreases in plant growth, but the reasons for this are not known. The effects of long-term exposure of soils to γ-radiation have received little attention compared to studies of the effects of acute treatments.

When considering the effects of pollution of soils with radionuclides, it is important to distinguish between the effects of the radionuclide itself and the associated radiation. For example, soil microorganisms are generally tolerant of plutonium, but its toxic effect is due to radiation. Leachate samples from disposal sites for low-level waste indicate that significant microbial activity occurs at these sites. Microorganisms may solubilise radionuclides such as insoluble plutonium dioxide by the production of chelates. The accumulation of ^{60}Co and ^{137}Cs by microorganisms followed by movement and eventual death of cells can lead to movement of radionuclides through soils. Our understanding of the ways in which radiation disturbs metabolic processes and upsets the balance between organisms in soil is far from complete.

5.6 Nitrate leaching

The use of N fertilisers has increased rapidly during the past 30 years, with most fertiliser N applied as NH_4^+-containing compounds such as NH_4NO_3 or urea. Urea, the most widely used N fertiliser worldwide, is rapidly hydrolysed to NH_4^+ in soil and then oxidised to NO_3^-. It is this form of N which is of greatest concern in terms of water quality.

The concern is that intensive use of N fertilisers may be responsible for the higher concentrations of NO_3^- recorded in surface and groundwaters. High concentrations of NO_3^- can cause eutrophication of lakes and streams as a result of excessive algal growth leading to depletion of O_2 as it undergoes decomposition. High concentrations of NO_3^- in leafy vegetables and in drinking water may also cause health problems if the NO_3^- is reduced to nitrite, although the medical evidence here is equivocal; NO_3^- itself is not toxic.

The maximum concentration of NO_3^- in drinking water recommended by the World Health Organisation is 50 mg l^{-1}, and this is also the maximum peak concentration allowed by the European Commission. Leachate from arable soils in the UK often exceeds this limit; measures are being introduced to place restrictions on agriculture in the catchment areas of water which either already exceeds this limit, or is at risk of exceeding the limit.

Studies in the UK have indicated annual rates of leaching of N from an unfertilised soil of 0.2 g m^{-2} under grass, 3.0 g m^{-2} under white clover and 13.7 g m^{-2} under bare fallow. Rates of leaching from fertilised grassland are higher under grazing than when grass is cut. The highest concentrations of NO_3^- in aquifers are found under fertilised arable soils, and large increases in concentrations occur following the ploughing of established grassland.

Data from the Rothamsted Drain Gauges established in 1870 indicate rates of NO_3^- leaching at the end of the nineteenth century from uncropped, uncultivated arable soils receiving no manure or fertiliser which are similar to current rates of leaching from fertilised wheat crops. From these data it can be concluded that the major source of NO_3^- leaching from arable soils is not directly from unused chemical N fertilisers, but from mineralisation of organic N (which may have been built up over a long period of time by fertiliser additions). The half-life for this organic N has been estimated to be 41 years, indicating the long-term nature of the problem. There are no obvious ways of reducing the concentration of NO_3^- in groundwaters in the short term, other than treatment of the water before consumption.

Apart from leaching, losses of N occur due to volatilisation of NH_3 under alkaline conditions, and denitrification. Emissions of NH_3 from animal slurry produced by intensive livestock enterprises are contributing to acid deposition in countries such as The Netherlands. These processes also contribute to the inefficiency of utilisation of N in both inorganic and organic fertilisers. Control over rates of urea hydrolysis and nitrification in soils offers one approach to reducing the losses of N from both inorganic and organic N fertilisers to groundwaters and to the atmosphere.

Table 5.3 The effect of nitrification inhibitors on the response of a range of crops to animal slurry (surface applied (S), or injected (I)). After B.F. Pain *et al.* (1987), in *Animal Manures on Grassland and Fodder Crops. Fertilizer or Waste*, eds. H.G. van der Meer *et al.*, Martinus Nijhoff, Dordrecht, pp. 229–246

Crop	Country	Slurry	Inhibitor	Yield ($g\ m^{-2}$)
Grass	UK	Cattle (S)	− DCD	220*
Grass	UK	Cattle (S)	+ DCD	300*
Maize	USA	Pig (I)	− Nitrapyrin	730
Maize	USA	Pig (I)	+ Nitrapyrin	1140
Sugar beet	The Netherlands	Pig (S)	−DCD	6200
Sugar beet	The Netherlands	Pig (S)	+DCD	7540

*First silage cut only

5.6.1 *Inhibitors of N transformations*

Numerous compounds have been patented as inhibitors of urea hydrolysis and of nitrification, but most are not particularly effective. The most interesting urease inhibitor is phenylphosphorodiamidate (PPD) but the activity of this compound is reduced at temperatures above 10°C. There appears to be scope for selecting similar compounds which are less affected by high temperatures and which do not affect other soil N transformations.

The control of nitrification appears more promising. Of the numerous patented compounds, three have attracted significant attention. These are 2-chloro-6-(trichloromethyl) pyridine (marketed as Nitrapyrin or N-serve), dicyandiamide (marketed as Didin or DCD) and 5-ethoxy-3-(trichloromethyl)-1,2,4-thiadiazole (marketed as Dwell or Terrazole). DCD breaks down in soil to form NH_4^+, NO_3^-, water and CO_2; however, the fate of nitrapyrin is uncertain and toxic residues may be produced. These compounds are useful in reducing the losses of N from animal slurries applied to the land during the winter when up to 20% of the total N applied may be lost by denitrification from injected slurry. Table 5.3 shows the effectiveness of Nitrapyrin and DCD in controlling the losses of N and hence increasing the yield of a range of crops receiving animal slurry.

5.7 Pesticides

Pesticides are currently an integral part of modern farming practice, contributing to increased agricultural productivity, and are being used increasingly in developing countries where pests and diseases are the major factors limiting food production. They include herbicides, insecticides, fungicides and nematicides.

Chemical pesticides, particularly naturally occurring products such as sulphur and mercury compounds, have been used for centuries. However, the use of synthetic organic pesticides originated in the 1930s with the discoveries of the insecticidal action of DDT and of the synthetic plant hormone herbicides MCPA and 2,4-D. The range of products has expanded dramatically over the last 50 years and in 1987 approximately 600 active ingredients were available to farmers around the world. In 1986, 3.5×10^6 tonnes of pesticides were used with a value of almost \$$16 \times 10^9$.

5.7.1 Side effects

Pesticides are by definition biologically active compounds, but it has not yet been possible to achieve absolute specificity, therefore non-target organisms are affected. As little as 1% of the pesticide applied may reach the target, and in the developed world a single crop may receive one or more applications of several pesticides each season. Countries vary in their regulatory requirements concerning the use of pesticides. Pesticide degradation and mobility in soil, together with aquatic toxicology tests, are usually required, and some countries require data on the effects of pesticides on soil microorganisms.

Pesticides can enter soil by a variety of routes (Figure 5.3). Apart from accidental spillage of chemicals, there may be overspraying or runoff from plants into soil. Faeces of treated animals and death of treated organisms lead to incorporation of compounds into soil. Pesticides may also move from soil to other parts of the environment, for example by aerial transport of soil, plant debris and spores, and by volatilisation, leaching and runoff from soil. Pesticides normally enter the soil at or below the concentrations recommended for agricultural use (0.05–0.5 g m^{-2}), which, assuming a soil bulk density of 1.0 g cm^{-3} and an even distribution within the top 10 cm of soil, would give a concentration in soil of 0.5–5 mg kg^{-1}.

A compound such as DDT (Figure 5.4a) contains bonds such as C–Cl which are very strong and render the molecule resistant to degradation. Furthermore, it has low solubility in water but is lipophilic (dissolves in fat), and therefore accumulates in plants and animals. When it was first used, DDT was extremely effective at eradicating the insect vectors of diseases such as malaria and yellow fever, but excessive use led to the development of DDT-resistance in target insects. Consequently, many countries have now banned the use of this insecticide.

Insecticides may affect non-target soil animals, but the effects are generally minimal and often transient. Earthworms are not particularly susceptible to insecticides, but they can concentrate organochlorine compounds by a factor of up to nine. Different groups of organisms may be affected differently, and this may upset the interaction between predator and prey. For example, DDT causes a decrease in the population of predatory mites which leads to an increase in the population of springtails.

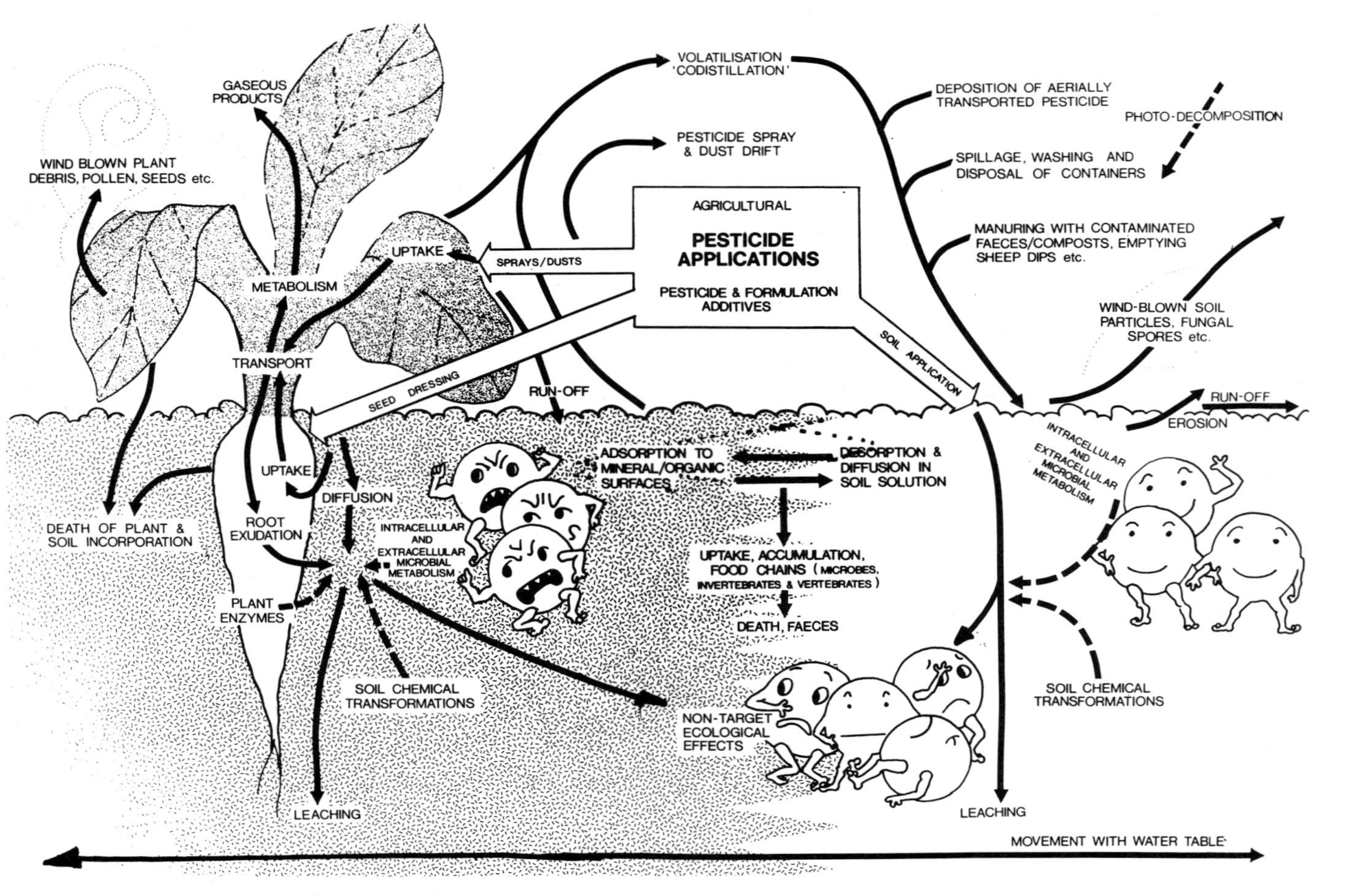

Figure 5.3 Pesticides in the soil environment (solid lines, movement of pesticides or products; broken lines, agencies affecting the pesticide or products; shaded area, rhizosphere; animations, microorganisms). Courtesy of Dr I.R. Hill, ICI.

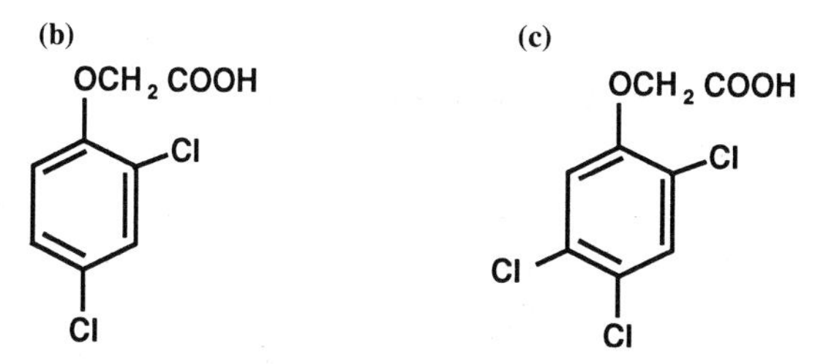

Figure 5.4 Structure of three pesticides. (a) the persistent insecticide DDT (1,1,1-trichloro-2,2-bis(*p*-chlorophenyl)-ethane); (b) the non-persistent herbicide 2,4-D(2,4-dichlorophenoxyacetic acid); and (c) the persistent herbicide 2,4,5-T (2,4,5-trichlorophenoxyacetic acid).

However, both mites and springtails are susceptible to nematicides (used to control nematode diseases). There is little information on the effect of fungicides on soil invertebrates.

In temperate cultivated soils the indirect effects of pesticides in allowing different tillage practices may have greater long-term impact on the soil invertebrate community than any direct effects of the pesticides. The introduction of hormone herbicides such as 2,4-D into cereal farming eliminated broad-leaved weeds, thereby allowing the continuous cultivation of wheat and minimum tillage systems. In this case the use of herbicides has indirectly increased the numbers of earthworms in soils under reduced cultivation.

Broad-spectrum biocides including soil fumigants often cause an increase in the growth of plants even when no obvious pests or diseases are present. This may involve the release of N from the pesticide or from the killed biomass, or the control of deleterious rhizobacteria perhaps by encouraging colonisation of soil by fluorescent pseudomonads, which often occurs after fumigation.

A change in the input of plant material into soil will have a major effect on all organisms. The rhizosphere is an important site for microbial activity, but little is known of the effects of pesticides on microorganisms in this zone of soil. Herbicides increase the amount of material lost from the roots, and plants exposed to herbicides may be more susceptible to plant disease than untreated plants. If herbicides do not penetrate more than a few centimetres into the soil then side effects in this zone may be unimportant when there is a large volume of deeper soil in which roots are present and microbial activity is proceeding normally. Herbicides do not appear to have any great or prolonged direct effect on the total bacterial population of the soil.

Despite many years of study it is not yet possible to unequivocally identify and quantify the effects of pesticides on soil organisms and on soil fertility. This is partly due to our lack of complete understanding of the role of organisms in the development and functioning of soil. Suitable techniques are also lacking. For example, the importance of symbiotic N_2 fixation in maintaining or improving the fertility of soils is well established, however, there are no completely reliable techniques available for routinely measuring rates of N_2 fixation by legumes and actinorhizal plants.

There is also a need to assess the significance of fluctuations in soil organisms and biological processes caused by a pesticide in relation to natural fluctuations. For example, drying and wetting of soil can lead to a 50% decrease in microbial activity, and organisms may take up to 30 days to recover. Pesticide effects which do not exceed these limits should not therefore present any risks.

One of the few data sets for the effects of a pesticide on bacteria in the field is shown in Figure 5.5. The application of methyl bromide, a fumigant, to a soil in South Australia used to grow wheat caused a large decrease in the bacterial population. However, within about 20 days the population increased to greater than its original size, and only after about

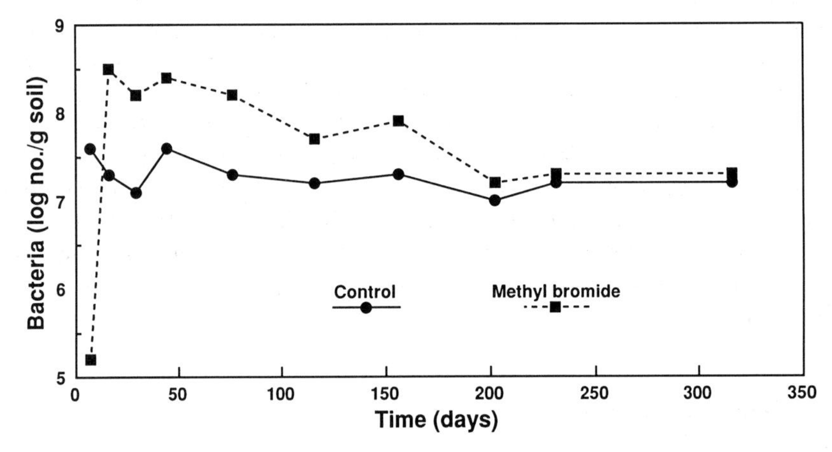

Figure 5.5 Changes in numbers of bacteria in a solonised brown soil (South Australia) in the field after fumigation with methyl bromide. After J.N. Ladd *et al.* (1976) *Soil Biology and Biochemistry* **8**, 255–260.

200 days did it revert to its original size. Was this effect of the soil fumigant ecologically significant?

5.7.2 Recalcitrant compounds

The magnitude of the side effects associated with a particular pesticide will depend partly upon the persistence of the compound in soil. The organochlorine insecticide DDT (Figure 5.4a) and the organo-phosphorus insecticide parathion (*O*,*O*-diethyl *o*-*p*-nitrophenylphosphoro-thioate) persist in soil for more than 15 years, whereas the herbicide 2,4-D (Figure 5.4b) disappears from soil within several weeks.

Persistence of a compound is determined largely by the ability of the soil microbial population to degrade the compound. The ability of plants to metabolise pesticides is also a factor. The reasons why some compounds are more resistant to attack by microorganisms (recalcitrant) in soil are complex. Simple structural changes can convert a rapidly decomposable substrate into a persistent compound. For example, the addition of an extra chlorine atom to a molecule of 2,4-D (Figure 5.4b) converts it into 2,4,5-T which persists in soil for several months.

Soils may develop the ability to degrade a pesticide following exposure to the compound. This apparent adaptation by the soil to compounds such as 2,4-D may involve induction of enzymes by microorganisms or the transfer of plasmids containing genes responsible for enzymes involved in the breakdown of the compound. For example, the genes responsible for degradation of 2,4-D are carried on a conjugative plasmid in *Alcaligenes paradoxus*.

Non-enzymatic reactions such as changes in pH brought about by microorganisms may also contribute to pesticide degradation. Similarly, recalcitrance may be partly due to adverse physical conditions such as low concentrations of O_2, or physical protection of the compounds by soil components. The metabolism of some pesticides by microorganisms may be as a result of cometabolism; the pesticide undergoes catabolism along with some other organic substrates almost as an incidental process. The supply of organic matter can therefore be an important factor in the degradation of compounds such as atrazine. It is likely that the cooperative action of communities of microorganisms is also important in the degradation of such xenobiotic compounds (chemicals that are foreign to the soil).

5.8 Introduced organisms

Microorganisms can be introduced into soil either indirectly as a side effect of the addition of organic materials to soil (e.g. animal slurry or sewage sludge) or directly as an inoculant introduced to the soil for a specific purpose. Concern over the fate of introduced organisms has been aroused particularly with the development of genetically modified organisms or GMOs (formerly, when restricted to microbes, knowns as GEMs).

If an important organism is absent from a soil then it may be necessary to introduce or inoculate that organism into the soil. For example, some trees will not grow unless the appropriate ectomycorrhizas are present, and the rhizobia present in a soil may not nodulate a new legume which has not previously been grown in that soil. Of course inoculation may be carried out by simply transferring soil from a region where the plant grows successfully to the new region.

However, as soil microbiological processes have become better understood attempts have been made to inoculate plants with pure cultures (or known mixtures of pure cultures). Such attempts to introduce an organism into an otherwise vacant niche are generally successful. However, inoculation may also be used in an attempt to supplant an indigenous organism, which is of poor quality, with a superior strain, or to suppress a troublesome organism, e.g. a pathogen. This is more difficult to achieve, and there are only a few reports of success.

Azotobacter chrococcum and *Bacillus megatherium* were used as seed inoculants in the former USSR and Eastern Europe for several decades. This bacterisation of crops was aimed at exploiting the N_2-fixing ability and phosphate solubilising abilities of *A. chrococcum* and *B. megatherium*, respectively. Yield increases of up to 10% were achieved with *A. chrococcum*, but the beneficial effects were probably due to the excretion of plant growth hormones. The responses of inoculation with *B. megatherium* were less convincing.

Organisms may also be introduced to remove or overcome the effects of a detrimental organism such as a pathogen (biological control). Such uses of biological agents to improve plant growth may be considered more desirable than the use of chemical means, particularly in cases where chemical treatment is not particularly effective. However, as the techniques of molecular biology are applied to the development of microbial inoculants, the acceptabilty of biological agents is being called into question.

5.8.1 Introductions of non-engineered organisms

There is a large body of information on the fate of non-engineered microorganisms, in particular *Rhizobium* sp. and *Bradyrhizobium* sp., introduced into soil. *Rhizobium* inoculants have been used for almost 100 years and millions of hectares are inoculated annually. There have been no unforeseen effects associated with this massive soil inoculation programme, despite the fact that many of the inoculants used may have been contaminated with organisms other than *Rhizobium*. On the contrary, the evidence from *Rhizobium* inoculation suggests that it is extremely unlikely that an introduced organism will persist in soil unless there is a vacant niche for the introduced strain.

One of the organisms of interest as a soil inoculant is *Pseudomonas* sp., a genus which is similar in many respects to *Rhizobium*. Studies on *Rhizobium* have indicated that survival of inoculants in soil depends upon a wide range of soil factors such as acidity, temperature, moisture content, predation and antagonism from other soil organisms. The response of *Pseudomonas fluorescens* to acidity and related factors is similar to that of *Rhizobium*, therefore the information available on root nodule bacteria should be valuable in predicting the fate of pseudomonads in soil.

Large numbers of microorganisms are introduced into soil when animal slurry is applied to the land. The microbial population in slurry is variable, but contains a proportion (approximately 1%) of coliform bacteria, which are normally found in the guts of animals and some of which may be pathogenic to humans. Coliforms do not usually survive more than a few weeks in the soil, but this could be long enough to allow genetic transfer to occur.

Escherichia coli, a coliform bacterium, is widely used in recombinant DNA studies. It can transfer genes to about 40 genera of Gram-negative bacteria including some pathogens. In this type of work strains which are unable to colonise the human gut, and which are unable to survive in soil are used.

5.8.2 Genetically modified organisms

The genetic material in a bacterium is contained in the chromosome, a large circular piece of DNA containing approximately 1000 kb (the average gene contains 900 base pairs, 1 kb equals 1000 base pairs, and 1 MDa equals approximately 1.5 kb), and plasmids which are smaller circular pieces of extrachromosomal DNA containing 1–300 kb. Plasmids, which can carry a substantial proportion of a cell's genetic material and which may be present in multiple copies (up to 40), replicate independently of the chromosome and often carry genes which assist with survival under adverse conditions. For example, resistance to antibiotics and to heavy metals, and catabolism of certain sugars and xenobiotics, together with the ability to form root nodules and crown galls are all determined by plasmid-borne genes.

Transposon mutagenesis is often used to locate particular genes in a bacterium. A transposon is a highly mobile short piece of DNA, often containing antibiotic-resistance genes as markers, which integrates itself at random into the chromosome or plasmid, and in the process deletes genes and causes a mutation. Mutants are selected which have lost the particular gene(s) of interest, and the position of the transposon and hence the original gene(s) can be located using DNA hybridisation techniques.

There are three main ways in which DNA may be transferred between bacteria of the same or different species. Conjugation, which only occurs in Gram-negative bacteria, involves conjugative plasmids (a minority of plasmids) and requires cell-to-cell contact between donor and recipient during which the plasmid and all or part of the chromosome is transferrred. In transduction a temperate bacteriophage (a virus) adsorbs to the surface of a bacterium and injects its viral DNA into the bacterial cell where it may integrate into the chromosome or exists as a plasmid. Death of the bacterial cell does not occur until the cell is exposed to agents such as ultraviolet light which induce the cell to produce intact phage particles which may contain additional DNA from the bacterium. The cell then lyses and the phages are released and may infect other bacteria. Transformation occurs under adverse conditions when bacteria can take up naked DNA which becomes integrated into the host chromosome.

Recombinant DNA technology offers the possibility of improving microbial inoculants for use in soils. When suitable genes have been identified they can be introduced into a vector such as a plasmid or a bacteriophage. This involves cutting and joining the vector DNA. The genes are usually transferred into *E. coli* and amplified (cloned), before being transferred to the desired host organism. Some progress has already been made, for example *Agrobacterium rhizogenes* strain 84. Strains of *Rhizobium* sp. have been constructed with altered host range (Table 5.4) and with introduced antibiotic-resistance marker genes. Genes have been identified in *Pseudomonas fluorescens* which determine the production of a phenazine-type antibiotic which appears to be essential for the suppression of take-all disease.

The chitinase genes from *Serratia marcescens* have been cloned in *E. coli* and transferred to *P. fluorescens*, an efficient root-colonising organism,

Table 5.4 Nodulating ability of *Rhizobium leguminosarum* biovar transconjugants produced by *Sym* plasmid transfer. After M.A. Djordjevic *et al.* (1983) *Journal of Bacteriology* **156**, 1035–1045

Strain	Phenotype	Nodulation response on	
		White clover	Pea
ANU843	Wild type	Nod+	Nod−
ANU845	*Sym* plasmid cured	Nod−	Nod−
ANU845 (pBR1AN)	Transconjugant	Nod+	Nod−
ANU845 (pJB5JI)	Transconjugant	Nod−	Nod+
ANU300	Wild type	Nod−	Nod+
ANU615	*Sym* plasmid cured	Nod−	Nod−
ANU615 (pBR1AN)	Transconjugant	Nod+	Nod−
ANU615 (pJB5JI)	Transconjugant	Nod−	Nod+

and it is hoped that plants inoculated with this organism will be protected against fungi with chitinous cell walls and possibly soil-inhabiting insects. *Bacillus thuringiensis* is another organism of interest because it produces large insecticidal proteinaceous crystals which are lethal to Lepidoptera, Coleoptera and Diptera. Spores and crystal preparations from *B. thuringiensis* have been used for almost 35 years to control Lepidoptera. The toxins are attractive because they are active in low doses and are highly specific. The genes which allow production of these toxins have been inserted into the chromosome of a root-colonising fluorescent pseudomonad with the aim of using this organism to control cutworm.

Microorganisms in soil and water exhibit a wide range of degradative abilities; however, in order to exploit these abilities it may be necessary to accelerate the process or combine the abilities of several organisms into a single strain. Many xenobiotics are recalcitrant (section 5.7.2) and it is therefore desirable to develop new organisms to degrade them. This may involve the modification of an existing pathway for a similar compound, or the design of a new pathway. Microorganisms have been constructed which are capable of degrading chemical pollutants such as 2,4,5-T.

5.8.3 *Risk assessment*

The deliberate and accidental release of genetically modified organisms into soil raises several important questions which need to be considered before permission is given by the appropriate legislative bodies for the use of such an organism. Questions include the following:

Can the organism survive, multiply and migrate in soil?
Is the novel gene transferred to other organisms?
Does the GEM affect other soil processes?
What are the overall risks associated with the use of the organism?

The answers to these questions are not readily available. In order to predict the likely impact of an introduced organism in soil a thorough understanding is required of the ecology of the soil. Although we have some information on certain organisms in soil, and on how they interact with other organisms and with the soil, we are still far from a complete understanding. We still do not know what properties confer on a bacterium the ability to compete successfully with other soil organisms, or to colonise the surface of a root. At present it is impossible to predict with any degree of certainty the fate of an introduced organism in soil or to design a successful inoculant for use in the field. This is largely due to the problems in knowing the precise nature of the ecological niche into which the organism is to be introduced.

It is necessary therefore to carry out a case-by-case assessment of the risks associated with the use of a particular GMO. A GMO should be evaluated according to phenotype; precise genetic characterisation does not ensure that all ecologically important aspects of phenotype can be predicted for environments into which an organism will be introduced.

It is essential that reliable methods are available for detecting an introduced organisms in soil. A range of techniques is available including traditional ecological methods such as culturing on selective media and fluorescent antibody techniques, which allow the phenotype of the organism to be monitored. Recently developed techniques such as ^{32}P-labelled DNA probes, which only hybridise (attach) to complementary sequences of DNA, allow the genotype of the organism to be monitored. The sensitivity of these probes is greatly increased when used in conjunction with the polymerase chain reaction (PCR) which amplifies nucleic acid sequences.

Molecular markers such as antibiotic-resistance, *lacZY* genes and the use of the bioluminescence genes are also useful. Organisms expressing the *lacZY* genes, which code for β-galactosidase and lactose permease, may be detected by growth on solid medium containing the substrate X-Gal by the appearance of blue/green colonies as the substrate is cleaved. This is particularly useful for marking pseudomonads which do not normally possess *lacZY* genes. Bioluminescence-based marker systems involve the introduction of genes for light emission derived either from the bacterial luciferase system (*lux*) or from the firefly luciferase system (*luc*). This technique allows marked organisms to be located *in situ*; when used with charge-couple device (CCD) imaging this allows the detection and quantification of single photons.

Risks can be assessed in model soil systems in the laboratory (microcosms) in which field conditions can be simulated as closely as possible under strict containment. It may be desirable to simplify the system, for example by sterilising the soil, or by using sieved soil, however, the extrapolation of results from microcosms to the field becomes more difficult as microcosms become less representative of field conditions. No single detection technique can provide all the information required for risk assessment, and the combined use of traditional and molecular techniques is essential.

Transfer of genetic material between bacteria takes place in soil. Conjugation has been reported in sterile and non-sterile soil for *E. coli* and *P. aeruginosa*, transformation in sterile soil for *Bacillus subtilis*, and transduction in sterile soil for *E. coli*. The optimum temperature for conjugation is above 20°C and the optimum pH value greater than 6. Bacteria may lose plasmids at high temperatures, for example *Rhizobium* sp. lose the *Sym* plasmid when incubated at 37°C. Novel genes should be more stable if located on the chromosome rather than on plasmids. Sites of high population density in soil, such as the rhizosphere and invertebrate guts, may provide bacteria with greater opportunities for genetic exchange.

Subtle differences in properties such as antibiotic resistance and metabolic capacity between donor and recipient transconjugants may influence competitive ability of these organisms in soil, although this has not been demonstrated. There is some evidence that transconjugants formed in the laboratory are less able to survive in soil than the parent organisms. However, other evidence indicates no change in properties such as root-colonising ability.

So although there appears to be considerable potential for engineering microorganism to carry out specific tasks in soil, the lack of predictive information on the ecological consequences of introducing such organisms may preclude their general use. It is more likely that if this technology is to be put into practice then it will be under more contained conditions such as in glasshouses, and not in the wider environment, unless by accidental release.

6 Soil biology – into the next millennium

6.1 Introduction

There can be little doubt that the thin, delicate layer of soil that covers one-third of the surface of our planet plays a vital role in sustaining life on Earth. We may not have always appreciated this fact and afforded soil the respect it merits, but as we approach the end of the twentieth century, a century of extraordinary scientific and technological achievements, it is becoming clear that the continued survival of our civilisation depends even more than ever upon our relationship with the land and soil.

The bonds that formed over thousands of years of intimate association between our ancestors and the land, as source of food and fibre, have been strained during the last few centuries. Industrialisation and urbanisation have severed the link for many people; food and clothing are now provided by institutions; festivals, once a celebration of the bounty of nature, are now occasions almost exclusively for increased consumption.

Yet despite these trends, which have taken place only very recently in our long history, there are signs that a new awareness is growing of the importance of the environment around us and of the land, air and water which sustain us. These signs are evident in the debate which was most obviously stimulated (though not initiated) by the report in 1987 of the World Commission on Environment and Development. This report was instrumental in launching the concept of sustainability. A notable outcome of this was the unique meeting of political leaders from around the world at the United Nations Conference on Environment and Development at Rio de Janeiro in June 1992. A series of global initiatives emerged from this Earth Summit.

This final chapter examines several issues, particularly those which relate to the management of soil and land, which reflect the increasing awareness of the impact of human activity on our planet.

6.2 Sustainability – so what's new?

"Sustainability is not business as usual with a few concessions to environmental management. Sustainability is a challenge to confront new realities, to devise new ways of doing things . . .". This quotation comes from a symposium on Sustainable Land Management (Hayward and

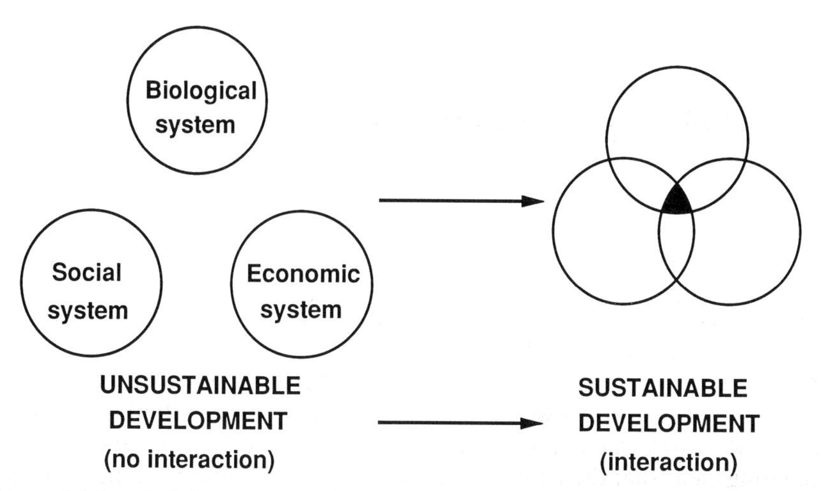

Figure 6.1 Sustainability as a trade-off between three interacting spheres of activity. After J. Holmberg and R. Sandbrook (1992) in *Policies for a Small Planet*, ed J. Holmberg, Earthscan Publications, London, pp. 19–38.

McChesney, 1992) held in New Zealand, a country which in 1991 introduced the Resource Management Act which uses the concept of sustainability as the basis for future management of natural resources. There are enormous implications here for the management of land and water.

Sustainabilty cannot of course be a fixed goal; if we use the analogy that we are borrowing the Earth from our children, then it becomes clear that we cannot anticipate the needs of future generations. What it does entail is a continuous trade-off between the three main components of our world which can be considered as biological, economic and cultural. Figure 6.1 shows a schematic representation of what is involved.

There are other aspects to sustainability: many of the issues are common to all inhabitants of the Earth, and therefore require concerted action. To this end the industrialised countries have a clear ethical and moral obligation towards the developing countries. Progress towards sustainability requires effective participation of all groups in society. Furthermore, the distinction between agricultural priorities and environmental effects becomes irrelevant; in the future a more holistic approach will be needed to all issues involving management of soil and land.

6.3 Environmental quality

Environmental quality is largely influenced by human activity in terms of, first, the exploitation of resources and second, the production of waste

(Figure 6.2). Environmental impact can be expressed as a function of several variables as follows:

$$\text{Impact} = \int \{\text{population size, affluence, technology}\}$$

Affluence can be considered here as per capita consumption. One of the important variables in this relationship is the size of the population. Recent trends are shown in Figure 6.3. The world population is set to double during the next 50 years; this has enormous implications in terms of food production, resource use and waste production. It is clearly going to be important to ensure that the predominantly linear process shown in Figure 6.2 is converted wherever possible into a cyclical process.

This is most obvious in the case of the nutrients produced from waste-water treatment and from animal production systems. These are currently viewed as waste materials rather than as a resource, and are disposed of accordingly. It has been estimated that the amounts of N, P and K in farmyard manure and animal slurry produced in the UK are equivalent to 37, 65 and 97% respectively of the amounts of those elements applied as inorganic fertilisers. Organic manures are variable in composition and responses by crops are often variable. However, as our understanding of the biology of soil develops, our ability to utilise natural resources more efficiently should improve.

Principle 15 of the Rio Declaration on Environment and Development states: "Where there are threats of serious or irreversible damage, lack of full scientific certainty shall not be used as a reason for postponing cost-effective measures to prevent environmental degradation". International environmental agreements are taking on increasingly more elements of this

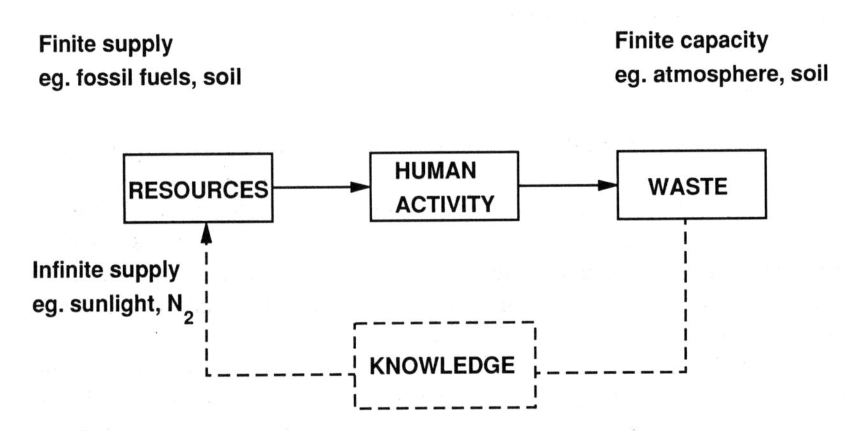

Figure 6.2 A schematic diagram of the flow of resources and waste products as a result of human activity. Solid lines indicate present pathways, dotted lines indicate pathways towards sustainability.

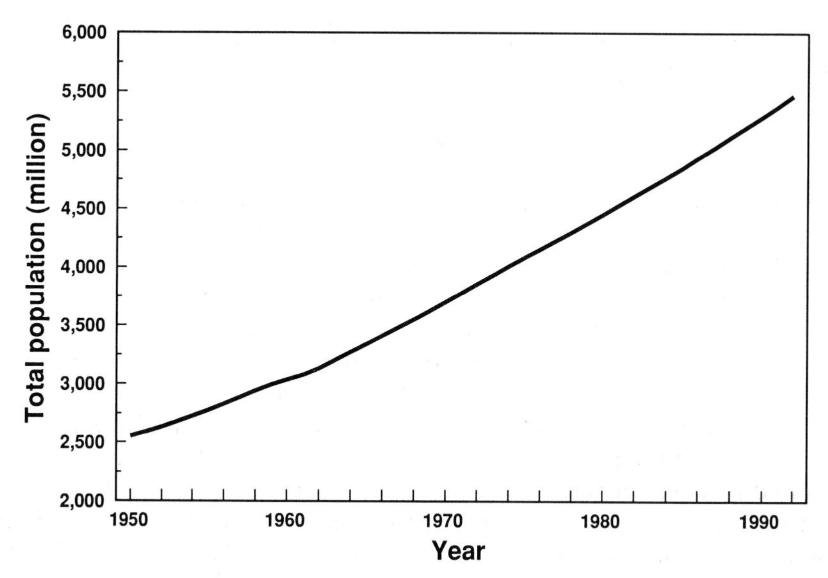

Figure 6.3 Trends in the world population 1950–1992. After L.E. Brown *et al.* (1993) *Vital Signs. The Trends that are Shaping our Future.* Earthscan, Publications, London.

principle. A good example of the application of the precautionary principle is in relation to global climate change (section 6.4).

A principle that is used in some current legislation is that of the 'polluter pays'. This is easier to implement in situations where the pollutant is derived from a point source e.g. a contaminated industrial site, but is impossible to implement where the pollutant is derived from a diffuse source e.g. nitrate contamination of groundwater.

6.3.1 Use of non-renewable resources

There is a concern that the polluter-pays principle may encourage the notion that pollution can be legitimised by counter-payment. It has been proposed that the concept of 'the user of natural resources pays' should be introduced to restrict the use of non-renewable resources. Such an approach would have a significant impact on an issue such as the relative merits of the use for food production of N fertiliser compared with N derived from biological nitrogen fixation.

During the 1970s estimates were obtained for the quantities of non-renewable energy required for the component processes of food production. Some interesting facts emerged: the energy required to produce and distribute a 1 kg loaf of sliced white bread is 21 MJ, almost the same as the energy required to produce 1 kg of sheep's fleece (enough to make a large woolly jumper) which is only 28 MJ. Compare these with the following data for production and distribution of different fertilisers:

1 kg N fertiliser	80 MJ
1 kg P fertiliser	16 MJ (as superphosphate)
1 kg lime	1 MJ

A comparison of the non-renewable energy input required to produce food in the United Kingdom and New Zealand (Table 6.1) shows that New Zealand is far more efficient. The reasons for this include the more favourable temperate climate in New Zealand with no extreme temperatures, and the greater reliance upon biological nitrogen fixation.

Inputs of phosphate and lime are required with legume-based pastures, but these are much less expensive in terms of non-renewable energy than is fertiliser-N. The legume root nodule, containing the enzyme nitrogenase, can be likened to a biological factory which obtains its energy from sunlight. In contrast, a N-fertiliser factory uses energy derived from fossil fuels. Using the data presented above we can compare the inputs of non-renewable energy to a legume-based pasture and to a pasture relying on N fertiliser.

If, irrespective of N source, annual inputs of P and lime to pastures are required of 20 kg ha^{-1} and 500 kg ha^{-1} respectively, and if root nodules can supply the equivalent of 200 kg N fertiliser per hectare, then the annual energy input to the two different systems is as follows:

(a) fertiliser-N-based pasture 16 820 MJ ha^{-1}
(b) nitrogen fixation-based pasture 820 MJ ha^{-1}.

So, clover-based pastures are twenty times more efficient in terms of non-renewable resources than fertiliser-N-based pastures.

If we accept that, first, reserves of fossil fuels are finite and second, any system of food production involves the removal of nutrients from the soil, then a system which relies only on inputs of P and lime must be more sustainable that a system which relies on these inputs but also requires the use of large amounts of N fertiliser.

It should also be pointed out that biological nitrogen fixation imposes no

Table 6.1 Energy efficiency of food production in the United Kingdom (UK) and New Zealand (NZ). After I.G. McChesney *et al.* (1982) *Energy in Agriculture* **1**, 141–153

Food	Energy (MJ per kg protein) UK	NZ
Milk	208	68
Wheat	45	25

burden compared to fertiliser N in terms of nutrient requirements (e.g. phosphate) or crop yield. It seems clear that irrespective of the wealth of a nation the use of biological nitrogen fixation should be encouraged in preference to the use of fertiliser N.

6.3.2 Soil quality

The aim of sustainable resource management programmes must be to maintain or improve the quality of soil in its broadest sense. The concept of soil quality is in many respects difficult to grasp; the criteria will depend heavily upon the use to which the soil is to be put. Soil scientists can collate series of chemical, physical and biological properties which are indicative of soil which conforms to a certain standard. However, soil and land also need to be considered from a holistic point of view, particularly with respect to the function of soil within the landscape. The notion of soil husbandry is a useful one in this context in relation to land management.

Contamination of soil and land is a specific concern where detailed guidelines are available for the maximum concentration of particular compounds or elements. Heavy metal contamination is one example, and Table 6.2 shows the current European guidelines.

In recent years large increases have been observed in the area of desert. Much of this land was formerly agricultural land and each year 2×10^5 km^2 deteriorate to the point of zero economic yield. Such a marked reduction in soil quality is due to over-cultivation, deforestation, over-grazing and unskilled irrigation. All of these factors, which are exacerbated by increasing population pressure, lead to a decrease in soil organic matter and increased susceptibilty to wind and water erosion.

The clearing and burning of tropical rain forest, which is rich in Ca and K relative to the soil, and its replacement by pasture or plantation forestry often leads to decreases in productivity within a few years. The reduced ground cover and lower inputs of organic matter allow erosion of the soil and further loss of nutrients.

Soil erosion can be viewed as the ultimate low point in terms of soil quality i.e. the soil is no longer there. Erosion of soil by wind and by water are major global problems. The dust bowl in the USA in the 1930s was a major disaster, brought about by a combination of climatic factors and poor soil management. As farmers in developing countries are forced to

Table 6.2 Maximum metal concentrations in soils (mg per kg of soil) permitted under EC regulations. After A. Wild (1993) *Soils and the Environment: An Introduction*, Cambridge University Press, Cambridge

Metal	Cd	Zn	Cu	Pb	Ni
Maximum concentration	3	300	140	300	75

bring steeper land into cultivation the risks of soil erosion become greater. There are simple technologies available such as contour ploughing and the construction of terraces, which can have a major impact on reducing this problem. A detailed discussion of this important topic cannot be given here, but there is an abundant literature.

6.3.3 Soils, subsoils and water quality

The role of soils in affecting the composition of water supplies, particularly in relation to drinking quality, has become a major issue. Within Europe this was stimulated by the 1980 Directive from the European Commission on the quality of water for human consumption which established the maximum admissible concentration (MAC) for nitrate and for pesticides in drinking water).

The following limits were set on the basis of amounts allowed per litre (dm^{-3}) of water:

(a) nitrate 50 mg as NO_3^- (which is equivalent to 11.3 mg N);
(b) any single pesticide 0.1 μg (0.5 μg total pesticide).

It is worth noting that the MAC for a pesticide is 500 000 times lower than the MAC for nitrate. Whereas the latter is based upon a certain amount of scientific evidence, the pesticide MAC is not based upon toxicological criteria (section 6.5.3), but is an effective zero. Several herbicides have been detected in groundwater in Europe and in North America at concentrations close to, or above, the MAC.

To understand the fate of chemicals applied to the soil surface is a complicated task. Factors such as residence time or persistence (determined by the ability of the soil population to degrade the chemical), adsorption to soil particles, and in particular the movement of water through the soil will be important factors. It has been generally assumed that once a chemical moves below the soil layer into the subsoil or unsaturated zone, and eventually into the saturated zone, the potential for degradation by microorganisms becomes negligible. However, recent studies in North America and in Europe have demonstrated the existence of variable but significant populations (up to 10^6 per gram of sediment) of bacteria at depth.

Studies of the microbiology of the deep subsurface have been stimulated by the concern over contamination of groundwater supplies, particularly by organic contaminants from industrial operations. The practical difficulties in obtaining uncontaminated samples from perhaps several hundred metres below ground are considerable, but techniques have been successfully developed. One possibility is that subsurface microorganisms might be encouraged to degrade certain contaminants. This technique of bioremediation is an area of much current interest.

Attention in the USA has been focused on sites such as Hanford in south-eastern Washington State. This site was acquired by the US government in 1943 and used until 1960 for the production of plutonium for the atomic weapons programme. It is underlain by a thick unconfined aquifer into which large volumes of waste water have been discharged. The site is contaminated with radionuclides and industrial solvents such as trichloroethylene. Samples taken from the underlying sediment at depths of 173–197 m (estimated to be 6–8 million years old), contained bacterial populations of up to 10^3 per gram of sediment.

The concentration of dissolved organic carbon (DOC) is an important factor determining the distribution of microorganisms in the subsurface environment. DOC is the organic carbon passing through a 0.45 μm filter (the carbon which is retained on the filter is termed suspended or particulate organic matter). The concentration of DOC in soil water (interstitial water) is 2–30 mg dm^{-3}; the concentration decreases with depth and is normally less than 2 mg dm^{-3} in groundwater (a range of concentrations of 0.2–15 mg dm^{-3} has been reported). This decrease in DOC with depth is a result of chemical and biological processes in the soil. The low concentration of DOC in groundwater is also a result of its long residence time in the aquifer ranging from hundreds to thousands of years.

Organic contaminants comprise less than 5% of the DOC in groundwater, therefore do not make a significant gross contribution to energy and nutrient supply for subsurface microorganisms. However, organic contaminants have a major impact on the ecology of particular groups of organisms. The DOC of groundwater is derived from two main sources: surface organic matter or kerogen (the fossilised organic matter present in the geological material of the aquifer). In most cases of groundwaters with low DOC (0.2-1.0 mg dm^{-3}), the DOC probably originates from the action of bacteria on kerogen in the aquifer.

The origin of microorganisms in the deep subsurface is not known; it seems unlikely that they are survivors from diagenesis of the sediments, several million years ago. They could originate from organisms applied to the surface of the land. For example sewage sludge and animal slurry which are applied to land contain a large population of bacteria. Of greatest concern from an environmental health point of view is the presence of pathogenic viruses and bacteria in these materials.

Bacteria of intestinal origin are, for the main part, harmless; others such as *Streptococcus faecalis* are potential pathogens causing forms of food poisoning. More serious illnesses are caused by three major pathogenic organisms:

Vibrio cholerae (cholera)
Salmonella typhii (typhoid)
Salmonella paratyphii (paratyphoid)

Table 6.3 Retention of *Escherichia coli* in Kentucky soil cores (two soil types) either intact (macropores present) or sieved and mixed (macropores absent), based upon the ratio of the concentration of cells in the leachate compared to the concentration in the irrigation water ($c/c_O \times 100$). After M.S. Smith *et al.* (1985) *Journal of Environmental Quality* **14**, 87–91

	Retention	
Soil	Intact	Disturbed
Silt loam	78.0	99.8
Sandy loam	21.0	95.0

Although these organisms are substantially reduced in number by sewage treatment, they are not completely eliminated. Animal slurry, in contrast, is not normally treated before it is applied to the land. The survival time of pathogens deposited on land depends on factors such as exposure to sunlight. In the UK, *Salmonella* has been shown to persist on grass for 18 days to 24 weeks.

Leaching of microorganisms through soils, and possible contamination of surface water or groundwater, depends upon factors such as the morphology of the organism (including the presence or absence of flagella), adsorption to soil particles, and the type of pores and channels in the soil. The importance of macropores, formed by plant roots or earthworms, in providing rapid water-conducting pathways to the subsoil is illustrated by the data in Table 6.3.

Agricultural materials such as effluent from silage and animal slurry are also serious threats to water quality; the former is highly effective at removing oxygen dissolved in water (biological oxygen demand or BOD). It has been estimated that a small farm of 40 ha carrying a dairy herd of 40 cows, and 50 pigs has the potential pollution load equivalent to a village of 1000 inhabitants. Animal slurry may also contain pathogenic bacteria; it is recommended that slurry is stored for at least 1 month before spreading on grazing land, then the land should be left ungrazed for at least 6 weeks.

6.4 Soils and global climate change

The Earth's atmosphere plays a vital role in the life on this planet, particularly in relation to the incoming solar radiation. Oxygen, nitrogen and ozone absorb otherwise harmful ultraviolet radiation from the sun; various components (water vapour, methane, nitrous oxide, ozone and chlorofluorocarbons) trap the longer wavelength radiation emitted from the Earth's surface (the greenhouse effect) thereby maintaining the mean surface temperature at 15°C.

There are currently two major concerns. First, that gases such as chlorofluorocarbons (CFCs), which are man-made compounds, are diffusing from the troposphere into the stratosphere (somewhere between 10 and 60 km above the Earth's surface) and destroying ozone. Second, that the atmospheric concentration of CO_2, CH_4, CFCs and N_2O is increasing, thereby leading to an increase in mean global temperature, i.e. global warming. The lifetime of these so-called greenhouse gases in the atmosphere ranges from 10 years for CH_4 to 150 years for N_2O. This is therefore a problem which requires urgent attention.

The main cause of global warming is the emission of CO_2 from the burning of fossil fuels, but the oxidation of soil organic matter may also contribute significant amounts. If all the carbon in soil organic matter were released into the atmosphere as CO_2 the concentration in the atmosphere would double from its present concentration of 350 ppm (parts per million by volume) to around 700 ppm. This is an unrealistic scenario; what is more likely is that a fraction of this carbon will be released as land is brought into cultivation, and organic matter oxidised by soil organisms.

Soils are a major source of methane; the gas is produced by bacteria during anaerobic respiration which only occurs under strongly reducing conditions. Such conditions can occur in natural wetlands (bogs, swamps, tundra), in rice paddies, and landfill sites. The concentration of CH_4 in the atmosphere, which is about 1.7 ppm at present, appears to be increasing at a rate of about 0.9% per year; this is more rapidly than the concentration of CO_2, and is partly due to the increase in area of rice grown on flooded soils.

Nitrous oxide (N_2O) is produced from soils during the process of denitrification and also during nitrification; the atmospheric concentration at present is estimated to be 310 ppb (parts per billion by volume). There is great spatial and temporal variability in the emission of N_2O from soils, and daily N losses of 0.1–10 g m^{-2} have been reported. If the production of N_2O from N fertiliser is only 1.5% of that applied, the contribution is significant (the annual use of N fertiliser around the world is estimated to be 60 million tonnes). It is clear therefore that soils may be playing an important role in the process of global warming.

As with any perturbation in a natural ecosystem, the issue of global warming should be viewed against the natural fluctuations in global temperature which have occurred in the past. Climate has varied widely during the recent history of the Earth, ranging from the equable climate of the Cretaceous and Eocene, to the comparatively cold conditions of the ice ages. The evidence suggests that major shifts in climate can take place rapidly over time-scales as short as centuries or even decades as illustrated by the Little Ice Age (a cold period lasting from about AD 1250 to AD 1850). Since the end of the last Ice Age (about 10 000 years ago) the average global temperature has differed from the present by 1–2 degrees. Since 1900, when temperatures were first measured directly, the increase has been about 0.45 degrees. It is not clear however, whether this is directly due to the increased concentration of greenhouse gases.

It is difficult to predict the short-term response of climate to increased concentration of greenhouse gases, partly because of the possibility of feedback loops introducing unforeseen complications. The Intergovernmental Panel on Climate Change suggested in 1990 that the mean global temperature could rise during the next century by 0.2–0.5 degrees per decade; this assumes that there is a reduction in the emission of CFCs, but no changes in the emissions of other greenhouse gases. Any reduction in the latter would reduce the likely global warming. Associated with this global warming is a predicted doubling of the atmospheric concentration of CO_2 by the second half of the next century.

An increase in CO_2 concentration will lead to a warmer atmosphere, especially near to the Earth's surface, but the implications for water vapour (the dominant greenhouse gas) in the atmosphere are less clear. Increased cloud cover could lead to a cooler planet if it allows less UV radiation to enter the atmosphere. An increase in mean global temperature will have effects on soil organisms and the processes they carry out, but the precise effects cannot be predicted. The use of simulation models and the effect of temperature on organic matter dynamics have been discussed in chapter 4.

The greatest impact of global warming is likely to be in the north temperate region, particularly ecosystems in the tundra and boreal zone where there are large reserves of relatively active organic matter in soils. This is currently an active area of scientific research. Under such circumstances, it is prudent to adopt the precautionary principle and take anticipatory preventive action.

6.5 Biodiversity

The current interest in biodiversity can be viewed as a development of the conservation movement; the need for action was formalised at the Earth Summit by a legally binding convention to prevent the eradication of biologically diverse species. Each contracting party was required to develop a national strategy; to this end the UK government published in 1994 the Biodiversity Action Plan.

Biodiversity is one of the threads that lies within the concept of sustainability. It is clear that biodiversity should be maintained because future practical needs and values cannot be predicted. Also, our understanding of ecosystems such as soil is not sufficiently well developed to be certain of the impact of deleting any component. Although it is believed that generally in microbial communities redundancy of function is common, this would clearly not be the case for nitrifying bacteria.

Early work on biodiversity equated diversity with the number of species present, but it is now generally accepted that there are at least three levels of study. We can distinguish genetic diversity (within species), organismal diversity (number of species) and ecological diversity (community diversity). When biological diversity is placed within the context of a physical environment, then we have an ecosystem. Chapter 2 provides an overview of our current knowledge of soil biodiversity.

The question of how we can measure biodiversity is a major challenge to scientists. It might be possible to count the number of species but, for an ecosystem such as soil where within only one gram there may be 10^9 individuals and up to 10^4 bacterial species, this is extremely difficult. In addition, not all species at a particular site may contribute equally to biodiversity. Molecular approaches such as nucleic acid homology or base sequence differences do not provide the solution; bacteria which have at least 70% DNA homology are considered to belong to the same species, but all hominids have at least 98% DNA homology. These kinds of inconsistencies and the discrepancies between DNA–DNA homology and 16s rRNA sequence data are troublesome. It appears that the goal of a universally accepted species concept remains as elusive as ever.

It has been suggested that biodiversity confers stability, i.e. the ability of a system to return to an equilibrium state after a temporary disturbance. It is known that many ecosystems can persist within varying physical environments such as seasonal climates (soil would be a good example of this). There are ecosystems, e.g. the savannah grasslands, which may need a disturbance such as fire to persist. Overall, it is not clear whether diverse systems resist disturbance better than simple ones (see also section 6.5.2).

The relationship between biodiversity and agriculture is an important one. Modern technological agricultural systems rely on a small number of plants and animals, and uniformity rather than diversity is the objective. This may be dangerous for two reasons: first, the objectives of technological agriculture and of conservation of biodiversity appear to be in conflict; second, simple biological systems may be unstable (in other words we are taking a major risk). However, as described in section 6.5, alternative agricultural methods based upon more biologically diverse systems are available.

The link between soil fertility and biodiversity can be considered further. It has been postulated that under favourable conditions biodiversity is low because the most effective competitors can dominate. Under poorer conditions such domination is not possible, and biodiversity is greater; under very poor conditions, however, few species can exist. It has been observed that plant biodiversity is greatest where soil fertility is lower, for example highly weathered tropical soils support diverse forest. When nutrients are added to such systems plant diversity decreases.

Furthermore, agricultural productivity is inherently high in areas where natural biodiversity is expected to be inherently low. Regions of high biodiversity such as tropical forest can only be used for highly productive agriculture if inputs of fertilisers etc. are used. This is not normally economically feasible, therefore biodiversity should be maintained. Such a hypothesis has important implications for the development of countries within tropical regions.

6.5.1 *Effects of management practice*

There are a few field data sets on the effects of soil and land management on soil microbial populations, although this has involved counting numbers of individuals within functional groups rather than numbers of species. More recently data have been obtained on changes in soil microbial biomass under different cropping systems, but these give no indication of biodiversity.

Early studies in the UK at Rothamsted examined the effects of additions of farmyard manure on bacterial populations. Samples taken from two plots at Broadbalk one of which had received manure for just over 100 years and one which was unmanured showed that the highest number of bacteria and actinomycetes was found in those plots which had received manure. Although there was no correlation between plate counts and direct counts of bacteria, both methods gave similar results.

In this study plate counts were positively correlated with temperature and direct counts were negatively correlated with temperature. Plate counts may provide an estimate of one component of the soil bacterial population which responds to external factors differently from the total population. However, data for soils with or without additions of farmyard manure indicated that an equilibrium existed between different components of the soil bacterial population. The equilibrium was not affected by these treatments despite differences in crop production.

A similar study in India using five different rice-based multiple cropping systems showed that the system of cropping used had little effect on the number of bacteria and fungi in soil. The seasonal effects, in particular temperature, had a more pronounced effect on the bacterial numbers.

6.5.2 *Soil resilience*

Soil, like all biological systems, undergoes continual change (see, for example, Figure 2.2), but nevertheless remains comparatively stable. This dynamic equilibrium is due in part to the cyclical nature of the biological processes: changes in one direction are accompanied by compensating changes in another direction. In well-established soil the microbial population seems to reach a very stable equilibrium, and although fluctuations and seasonal changes may occur in the abundance of individual species, the population as a whole is very resistant to change. This is most evident when attempts are made to introduce exotic species into the community (e.g. soil inoculants).

This characteristic of soils has been termed resilience; in the broadest sense it includes the buffering capacity of the soil in respect of physical, chemical and biological impacts. In relation to the biological quality of soil, resilience can be considered as the capacity of the soil population to maintain structure, diversity and function. This is obviously closely related to the concept of biodiversity.

6.5.3 *Ecotoxicology*

As our understanding of soil as a living system improves, more attention will need to be devoted to studying the injurious effects of chemicals and physical agents on soil organisms. This leads us into the field of toxicology, and in particularly ecotoxicology. The science of toxicology can be traced back to Paracelsus who in the sixteenth century postulated that "dosage alone determines poisoning".

Interest today is focused on the effects of biologically active compounds such as agrochemicals. There is also interest in the ecological relationships brought about by natural products, i.e. ecological toxicology. The use of any biologically active compound or element poses potential problems of toxicity to non-target organisms. Of course, if toxicologists did their job perfectly then there would be no problem. However, our knowledge of testing is imperfect; it is impossible to test all the species that may eventually prove to be affected to a significant degree. A particular compound may therefore be used under circumstances where there is a danger, but the danger is known. In other words a certain level of risk is accepted.

When considering the toxicity of compounds entering soil, there are two major considerations: first, how toxic is the compound to soil organisms? second, if that compound remains as a residue in food or if it enters the water supply how toxic is the compound to humans? The first question has been addressed in chapter 5, and the second was briefly considered in section 6.3.2.

Data are presented in Table 6.4 on the toxicity (as determined by the dose which causes 50% mortality in rats, i.e. the LD_{50}) of three herbicides and three common substances which we actively consume. A cup of coffee is clearly as toxic as many herbicides, and atrazine is no more toxic than table salt! It is estimated that 99.99% of dietary intake of toxins is from natural foods, and 0.01% is from synthetic pesticides.

Table 6.4 Toxicity of some herbicides and common chemicals to rats (acute oral LD_{50}, mg kg^{-1}) After A. Cobb (1992) *Herbicides and Plant Physiology*, Chapman and Hall, London; C.R. Worthing and R.J. Hance (1991) *The Pesticide Manual*, The British Crop Protection Council, Farnham, England

Compound	LD_{50}
Glyphosate	4320
Atrazine	2500
Paraquat	150
Table salt	3000
Caffeine	200
Nicotine	50

There are many reports of inhibitory or stimulatory interactions between plants, and between plants and microorganisms. For example, couchgrass (*Agropyron repens*), a weed found in many countries, decreases growth of wheat (*Triticum aestivum*) and lucerne (*Medicago sativa*). These types of interactions have been termed allelopathy, and it has been postulated that such interactions are mediated by chemicals termed allelochemicals (or allomones).

The adverse effect of the black walnut (*Juglans nigra*) on anything planted in its vicinity has been known since at least the time of Pliny in the first century AD. We now know that this is due to a toxic compound, juglone (Figure 6.4). A precursor of juglone is synthesised in walnut leaves, and when it is washed into the soil is rapidly hydrolysed to the allelochemical.

Another example comes from the UK where it has been observed that heather (*Calluna vulgaris*) reduces growth of Sitka spruce (*Picea sitchensis*) but not Scots pine (*Pinus sylvestris*). Also, concentrations of NO_3^- in cropped soil are often lower than in uncropped soil, even after taking into account the N taken up by the crop plants and leaching losses (which are lower under cropped soils than fallow soils). There is some evidence that nitrification is affected by allelochemicals produced by many plant species and microorganisms.

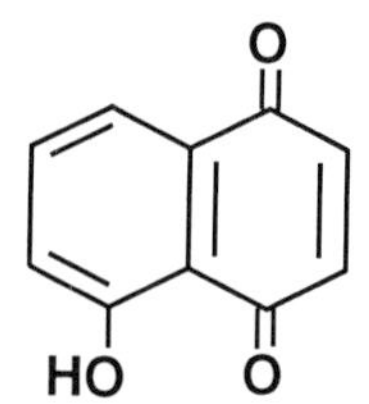

Figure 6.4 Juglone (5-hydroxy-1,4-naphthoquinone) a natural toxin produced in soil under walnut trees.

The mechanisms for many of these allelopathic interactions are not understood, but it is clear that the interactions between xenobiotic chemicals and soil organisms must be viewed against the background of an extraodinarily complex series of naturally occurring interactions. There may also be scope for managing allelochemicals to our advantage.

6.6 Sustainable agricultural systems

The remarkable increase in food production achieved in developed countries during the last 30 years, and more recently in some developing countries such as India, has been due to the breeding of high-yielding varieties of crops such as wheat (*Triticum aestivum*), increasing mechanisation and the use of large quantities of N and P fertilisers and pesticides. The major discoveries behind these developments were all made at least 50 years ago.

Surplus food production in developed countries together with increasing concern about the effects of agricultural practices on the environment are now prompting a critical evaluation of the long-term feasibility of high-input agriculture. At the same time efforts are continuing to improve food production in developing countries.

It is worth considering the implications of the projected increase in the world population (about 100 million new mouths to feed every year) in terms of food production. How much food does a person need? If we express a person's food needs in terms of a requirement for energy then it is reckoned that the absolute minimum for survival is 2600 kcal per day. This is equivalent to 286 kg of grain per person per year. Table 6.5 shows how much food people actually consume in a range of different countries.

These data show that the supply of food in Bangladesh and Kenya does

Table 6.5 Average grain consumption per person in 1990 for six countries (these data include grain consumed directly and that fed to animals and consumed as animal products). After L.E. Brown *et al.* (1994) *State of the World 1994, A Worldwatch Institute Report of Progress Towards a Sustainable Society*, Earthscan Publications, London

Country	Grain consumption (kg per person)
USA	860
Australia	503
China	292
Brazil	277
Bangladesh	176
Kenya	145

not meet the needs of the people; it also shows the delicate balance in two of the most heavily populated countries of the world, China and Brazil. The outlook is particularly bleak for Kenya; while it ranks much lower than Bangladesh in terms of people per hectare, if land is weighted by its potential productivity, then Kenya comes out with the highest population per unit of potential production. This means that it will face the greatest difficulty in attempting to feed its population.

Many of the improvements which contributed to the so-called green revolution during the 1970s were based on improved varieties of rice which responded well to inputs of fertiliser and pesticides. This was of benefit to the larger, more mechanised farmer, but did not help the smaller farmers. For example, only a small proportion of farms in south-east Asia have irrigation facilities and are able to take full advantage of the improvements in rice production by growing up to four crops per year. The majority of farmers cannot afford to buy fertilisers and are restricted to growing only one crop per year.

The traditional shifting cultivation (slash and burn) practised widely in the humid and subhumid tropics allows sustainable food production at low population densities. Following clearance of the forest, soil nutrients are depleted during a short cropping cycle and then restored during a fallow period. However, increasing population pressure has led to a decrease in the length of the fallow period with a consequent decrease in fertility and crop yields. Attention is therefore being focused on alternative low-input sustainable systems.

Mixed cropping is commonly practised in the tropics, often as an insurance against failure of one of the crops. This may involve intercropping (growing two or more crops simultaneously) or sequential cropping (growing two or more crops in sequence). Benefits are particularly pronounced when one of the crops is a legume. For example, the grain yield of a non-legume following a grain legume is often 50–100% greater than the yield of the same crop grown as a monoculture, equivalent to a fertiliser application (N) of 5–10 g m^{-2}.

The main contribution of the legume is its ability to fix N_2 thereby sparing some soil N for the non-legume. It is less likely that legumes provide an input of fixed N into the soil because the removal of N in the seed is often greater than estimated rates of N_2 fixation. Other benefits of having a legume in a mixed cropping system include improvements in soil structure, disruption of pest and disease cycles and an input of crop residues with a low C:N ratio which are more likely to result in net mineralisation of N during the early stages of decomposition than are crop residues from a non-legume (see also section 6.3.1).

The use of a woody species as a component of a mixed cropping system (agroforestry) is attractive because trees, particularly leguminous species, can improve recycling of nutrients and water from deep in the soil, provide shade and supply food, firewood and wood products. Most attention has been focused on *Leucaena leucocephala*, but other tree legumes such as *Sesbania* sp., *Prosopis* sp. and *Cassia* sp. may have greater potential. A system which uses pruned hedgerows of woody plants intercropped with food crops (alley cropping) has recently been developed.

Sustainable yields of crops have been achieved in the traditional home gardens and village forest gardens of Malaysia, Indonesia and the West Indies. These systems simulate the forest environment and have a multi-storeyed structure and a high diversity of cultivated species. On the island of Java mixed gardens have been in use since at least the tenth century and are still widespread in areas of high population density. A typical garden might include a ground layer of vegetables such as beans and tomatoes, a storey 1.5–5 m high of food plants such as cassava and papaya, and upper storeys with a range of different tree heights, for example citrus, coffee, mango, bamboo and coconuts extending up to 35 m. The complex interactions that occur between such a mixture of plant species are worthy of study.

6.7 Man and the Earth

Soil science as a subject is less than 100 years old, yet for the past 10 000 years it has been our knowledge of the land that has allowed the species *Homo sapiens* to flourish. As we move towards the millennium we are facing problems of unprecedented magnitude: such as how to feed an ever-increasing world population, how to protect the Earth's atmosphere. These are not solely environmental problems, they are issues which involve the fundamental relationship between human beings and the Earth.

It is clear that these problems will not be solved by scientists alone, but scientists will have an important role to play both as researchers and as teachers. Despite the ever-increasing trend towards specialisation, we shall need to draw on all available knowledge of land and soils. There is a role for everyone to play in the future of our planet.

There is much scientific work that remains to be done to improve our understanding of the biology of soil. The development of food production systems which exploit biological processes to a greater extent should now be the common aim of scientists in both developed and developing countries. However, the use of pesticides and inorganic fertilisers is likely to continue, combined with biological control agents, to form integrated systems of pest and disease control. Such strategies will involve a shift away from energy-intensive technologies to more knowledge-intensive technologies.

Several of the current objectives of modern agricultural research are compatible with the aims of those involved in organic farming (or biological agriculture). Biological agriculture aspires to a system in which the maintenance of soil fertility and the control of pests and diseases are achieved by the enhancement of natural processes and cycles, with only moderate inputs of energy and resources, while maintaining reasonable productivity.

The use of organic materials helps to maintain levels of soil organic matter, to improve soil structure and provides a source of nutrients for crop growth. Earthworm populations are also increased. For example, in the long-term experiments at Rothamsted, the Broadbalk plots which have received farmyard manure (3500 g m^{-2} per year) since 1843 have earthworm populations of 89 per m^2 compared to populations of 6 per m^2 in the plots receiving no organic manure or fertiliser.

The biodynamic farming movement, which has developed out of a range of ideas initiated by Rudolf Steiner during the early part of this century, strives to develop a system of farming that includes ecological, economic and social aspects. This is a concept that is remarkably akin to that of sustainability.

In biodynamic farming the whole farm is viewed as a living organism. This view has been taken a stage further by James Lovelock in his Gaia hypothesis in which it is proposed that our planet is a living organism, a complex entity involving the biosphere, atmosphere, oceans and soil; the totality constitutes a cybernetic system which seeks to optimise the physical and chemical environment for life. This is an attractive, but as yet unproven hypothesis.

Finally, an image that may help to guide us through these challenging times is the one provided by Edgar Mitchell, one of the astronauts aboard Apollo 14 in 1971, who was privileged to see the extraordinary sight of the Earth rising above the moon; the experience was captured in the following words (Kelley, 1988):

> Suddenly from behind the rim of the moon, in long, slow-motion moments of immense majesty, there emerges a sparkling blue and white jewel, a light, delicate sky-blue sphere laced with slowly swirling veils of white, rising gradually like a small pearl in a thick sea of black mystery. It takes more than a moment to fully realise that this is Earth . . . home.

References and further reading

Chapter 1

Brady, N.C. (1990) *The Nature and Properties of Soils*, Macmillan, New York.
Cairns-Smith, A.G. (1985) *Seven Clues to the Origin of Life*, Cambridge University Press, Cambridge.
McLaren, R.G. and Cameron, K.C. (1990) *Soil Science. An Introduction to the Properties and Management of New Zealand Soils*, Oxford University Press, Auckland.
Rowell, D.L. (1994) *Soil Science Methods and Applications*, Longman Scientific and Technical, Harlow, UK.
Stevenson, F.J. (1982) *Humus Chemistry, Genesis, Composition, Reactions*, John Wiley and Sons, New York.
Stotzky, G. (1986) Influence of soil mineral colloids on metabolic processes, growth, adhesion, and ecology of microbes and viruses. In *Interactions of Soil Minerals with Natural Organics and Microbes*, eds. Huang P.M. and Schnitzer, M., Soil Science Society of America, Madison, pp. 305–428.
Theng, B.K.G., Tate, K.R., Sollins, P., Moris, N., Nadkarni, N. and Tate, R.L. (1989) Constituents of organic matter in temperate and tropical soils. In *Dynamics of Soil Organic Matter in Tropical Ecosystems*, eds. Coleman, D.C., Oades, J.M. and Uehara, G., NifTAL Project, University of Hawaii, Honolulu, pp. 5–32.
White, R.E. (1987) *Introduction to the Principles and Practice of Soil Science*, Blackwell Scientific Publications, Oxford.
Wild, A. (1988) *Russell's Soil Conditions and Plant Growth*, Longman Scientific and Technical, Harlow.
Wild, A. (1993) *Soils and the Environment: An Introduction*, Cambridge University Press, Cambridge.

Chapter 2

Alexander, M. (1977) *Introduction to Soil Microbiology*, John Wiley and Sons, New York.
Ayanaba, A. and Sanders, F.E. (1981) Microbiological factors. In *Characterization of Soils in Relation to their Classification and Management for Crop Production: Examples from some areas of the Humid Tropics*, ed. Greenland, D.J., Clarendon Press, Oxford, pp. 164–187.
Clarholm, M. (1984) Heterotrophic, free-living protozoas: neglected microorganisms with an important task in regulating bacterial populations. In *Current Perspectives in Microbial Ecology*, eds. Klug, M.J. and Reddy, C.A., American Society for Microbiology, pp. 321–326.
Clark, F.E. (1967) Bacteria in soil. In *Soil Biology*, eds. Burges, A. and Raw, F., Academic Press, New York, pp. 15–49.
Duboise, S.M., Moore, B.E., Sorber, C.A. and Sagik, B.P. (1979) Viruses in soil systems. *CRC Critical Reviews in Microbiology* **7**, 245–285.
Edwards, C.A. and Lofty, J.R. (1977) *Biology of Earthworms*, Chapman and Hall, London.
Freckman, D.W. and Caswell, E.P. (1985) The ecology of nematodes in agroecosystems. *Annual Review of Phytopathology* **23**, 275–296.
Gregory, P.J., Lake, J.V. and Rose, D.A. (1987) *Root Development and Function*, Cambridge University Press, Cambridge.
Griffin, D.M. (1972) *Ecology of Soil Fungi*, Chapman and Hall, London.

Harley, J.L. and Smith, S.E. (1983) *Mycorrhizal Symbiosis*, Academic Press, New York.
Kerr, A. (1987) The impact of molecular genetics on plant pathology. *Annual Review of Phytopathology* **25**, 87–110.
Kevan, D.K. McE. (1962) *Soil Animals*, H.F.G. Witherby Ltd., London.
Küster, E. (1967) The actinomycetes. In *Soil Biology*, eds. Burges, A. and Raw, F., Academic Press, London, pp. 111–127.
Moore, J.C., Walter, D.E. and Hunt, H.W. (1988) Arthropod regulation of micro- and mesobiota in below-ground detrital food webs. *Annual Review of Entomology* **33**, 419–439.
Rolfe, B.G. and Gresshoff, P.M. (1988) Genetic analysis of legume root nodule initiation. *Annual Review of Plant Physiology and Plant Molecular Biology* **39**, 297–319.
Russell, E.J. (1957) *The World of Soil*, Collins, London.
Tjepkema, J.D., Schwintzer, C.R. and Benson, D.R. (1986) Physiology of actinorhizal nodules. *Annual Review of Plant Physiology* **37**, 209–232.
Warcup, J.H. (1967) Soil fungi. In *Soil Biology*, eds. Burges, A. and Raw, F., Academic Press, New York, pp. 51–110.
Weaver, R.W. *et al.* (1994) *Methods of Soil Analysis Part 2: Microbiological and Biochemical Properties*, Soil Science Society of America, Madison.

Chapter 3

Bowen, G.D. and Rovira, A.D. (1976) Microbial colonisation of plant roots. *Annual Review of Phytopathology* **14**, 121–144.
Brock, T.D., Madigan, M.T., Martinko, J.M. and Parker, J. (1994) *Biology of Micro-organisms*, Prentice Hall, Englewood Cliffs, NJ.
Burns, R.G. (1977) *Soil Enzymes*, Academic Press, New York.
Cole, J.A. and Ferguson, S.J. (1988) *The Nitrogen and Sulphur Cycles*, Cambridge University Press, Cambridge.
Dindall, D.L. (1990) *Soil Biology Guide*, John Wiley and Sons, New York.
Jenkinson, D.S. and Ladd, J.N. (1981) Microbial biomass in soil: measurement and turnover. In *Soil Biochemistry*, Volume 5, eds. Paul, E.A. and Ladd, J.N., Marcel Dekker, New York, pp. 415–471.
Killham, K. (1994) *Soil Ecology*, Cambridge University Press, Cambridge.
Lynch, J.M. (1983) *Soil Biotechnology*, Blackwell Scientific Publications, Oxford.
Paul, E.A. and Clark, F.E. (1989) *Soil Microbiology and Biochemistry*, Academic Press, New York.
Ritz, K., Dighton, J. and Giller, K.E. (1994) *Beyond the Biomass. Compositional and Functional Analysis of Soil Microbial Communities*, John Wiley and Sons, Chichester.
Rovira, A.D., Foster, R.C. and Martin, J.K. (1979) Origin, nature and nomenclature of the organic materials in the rhizosphere. In *The Soil–Root Interface*, eds. Harley, J.L. and Scott-Russell, R., Academic Press, New York, pp. 1–4.
Sprent, J.I. (1987) *The Ecology of the Nitrogen Cycle*, Cambridge University Press, Cambridge.
Stanier, R.Y., Adelberg, E.A. and Ingraham, J.L. (1987) *General Microbiology*, Macmillan Education, Basingstoke.
Waksman, S.A. (1932) *Principles of Soil Microbiology*, Baillière, Tindall and Cox, London.
Williams, S.T. (1985) Oligotrophy in soil: fact or fiction? In *Bacteria in their Natural Environments*, eds. Fletcher, M. and Floodgate, G.D., Academic Press, New York, pp. 81–110.
Winogradsky, S. (1949) *Microbiologie du Sol. Problèmes et Méthodes*, Masson et Cie, Paris.
Wood, M. and McNeill, A.M. (1993) $^{15}N_2$ studies on nitrogen fixation by legumes and actinorhizals: theory and practice. *Plant and Soil* **155/156**, 329–332.

Chapter 4

Berthelin, J. (1983) Microbial weathering. In *Microbial Geochemistry*, ed. Krumbein, W.E., Blackwell Scientific Publications, Oxford, pp. 223–262.

Darwin, C. (1881) The formation of vegetable mould through the action of worms. In *The Essential Darwin*, ed. Ridley, M. (1987), Allen and Unwin, London, pp. 237–256.

Fyfe, W.S., Kronberg, B.I., Leonardos, O.H. and Olureunfemi, N. (1983) Global tectonics and agriculture: a geochemical perspective. *Agriculture, Ecosystems and Environment* **9**, 383–399.

Haynes, R.J. (1983) Soil acidification induced by leguminous crops. *Grass and Forage Science* **38**, 1–11.

Hillel, D.J. (1991) *Out of the Earth. Civilization and the life of the soil*, The Free Press, New York.

Jenkinson, D.S. (1981) The fate of plant and animal residues in soil. In *The Chemistry of Soil Processes*, ed. Greenland, D.J., John Wiley and Sons, Chichester, pp. 505–561.

Jenny, H. (1980) *The Soil Resource: Origin and Behaviour*, Springer-Verlag, Berlin.

Lynch, J.M. and Bragg, E. (1985) Microorganisms and soil aggregate stability. In *Advances in Soil Science*, Volume 2, ed. Stewart, B.A., Springer-Verlag, Berlin, pp. 133–171.

Miles, J. (1985) The pedogenic effects of different species and vegetation types and the implications of succession. *Journal of Soil Science* **36**, 571–584.

Molloy, L. (1993) *The Living Mantle. Soils in the New Zealand Landscape*, New Zealand Society of Soil Science, Lincoln, New Zealand.

Nye, P.H. and Tinker P.B. (1977) *Solute Movement in the Soil–Plant System*, Blackwell Scientific Publications, Oxford.

Passioura, J.B. (1988) Water transport in and to roots. *Annual Review of Plant Physiology and Plant Molecular Biology* **39**, 245–265.

Satchell, J.E. (1983) *Earthworm Ecology From Darwin to Vermiculture*, Chapman and Hall, London.

Schatz, A. (1963) Soil microorganisms and soil chelation. The pedogenic action of lichens and lichen acids. *Journal of Agricultural and Food Chemistry* **11**, 112–118.

Simonson, R.W. (1959) Outline of a generalized theory of soil genesis. *Soil Science Society of America Proceedings* **23**, 152–156.

Swift, M.J. and Boddy, L. (1984) Animal–microbial interactions in wood decomposition. In *Invertebrate–Microbial Interactions*, eds. Anderson, J.M., Rayner, A.D.M. and Walton, D.W.H., Cambridge University Press, Cambridge, pp. 89–131.

Swift, M.J., Heal, O.W. and Anderson, J.M. (1979) *Decomposition in Terrestrial Ecosystems*, Blackwell Scientific Publications, Oxford.

Wood, T.G. (1976) The role of termites (Isoptera) in decomposition processes. In *The Role of Terrestrial and Aquatic Organisms in Decomposition Processes*, eds. Anderson, J.M. and Macfayden, A., Blackwell Scientific Publications, Oxford, pp. 145–168.

Chapter 5

Addiscott, T.M., Whitmore, A.P. and Powlson, D.S. (1991) *Farming, Fertilizers and the Nitrate Problem*, C.A.B. International, Wallingford.

Alloway, B.J. (1995) *Heavy Metals in Soils*, 2nd edn, Blackie Academic and Professional, Glasgow.

Blank, L.W., Roberts, T.M. and Skeffington, R.A. (1988) New perspectives in forest decline. *Nature* **336**, 27–30.

Cawse, P.A. (1975) Microbiology and biochemistry of irradiated soils. In *Soil Biochemistry*, Vol. 3, eds. Paul, E.A. and McLaren, A.D., Marcel Dekker, New York, pp. 213–267.

Hassall, K.A. (1990) *The Biochemistry and Uses of Pesticides: Structure, Metabolism, Mode of Action and Uses in Crop Protection*, Macmillan, London.

Hopkin, S.P. (1989) *Ecophysiology of Metals in Terrestrial Invertebrates*, Elsevier Applied Science, London.
Nye, P.H. (1981) Changes in soil pH across the rhizosphere. *Plant and Soil* **61**, 7–26.
Prosser, J.I. (1994) Molecular marker systems for detection of genetically engineered microorganisms in the environment. *Microbiology* **140**, 5–17.
Rowbury, R.J., Armitage, J.P. and King, C. (1983) Movement, taxes and cellular interactions in the response of microorganisms to the natural environment. In *Microbes in their Natural Environments*, eds. Slater, J.H., Whittenbury, R. and Wimpenny, J.W.T., Cambridge University Press, Cambridge, pp. 299–350.
Rowell, D.L. (1988) The management of irrigated saline and sodic soils. In *Russell's Soil Conditions and Plant Growth*, ed. Wild, A., Longman Scientific and Technical, Harlow, pp. 927–951.
Somerville, L. and Greaves, M.P. (1987) *Pesticide Effects on Soil Microflora*, Taylor and Francis, London.
Sussman, M., Collins, C.H., Skinner, F.A. and Stewart-Tull, D.E. (1988) *The Release of Genetically-Engineered Micro-Organisms*, Academic Press, London.
Stotzky, G. and Babich, H. (1986) Survival of, and genetic transfer by, genetically-engineered bacteria in natural environments. *Advances in Applied Microbiology* **31**, 93–138.
Tiedje, J.M., Colwell, R.K., Grossman, Y.L., Hodson, R.E., Lenski, R.E., Mack, R.M. and Regal, P.J. (1989) The planned introduction of genetically engineered organisms: ecological considerations and recommendations. *Ecology* **70**, 298–315.
Tiller, K.G. (1989) Heavy metals in soils and their environmental significance, *Advances in Soil Science* **9**, 113–142.
Vreeland, R.H. (1987) Mechanisms of halotolerance in microorganisms. *CRC Critical Reviews in Microbiology* **14**, 311–356.
Warner, F.E. and Harrison, R.M. (1993) *Radioecology after Chernobyl: Biogeochemical Pathways of Artificial Radionuclides*, John Wiley and Sons, Chichester.
Wood, M. (1995) A mechanism of aluminium toxicity to soil bacteria and possible ecological implications, *Plant and Soil* **121**, 63–69.

Chapter 6

Alloway, B.J. and Ayres, D.C. (1993) *Chemical Principles of Environmental Pollution*, Blackie Academic and Professional, Glasgow.
Alexander, M. (1994) *Biodegradation and Bioremediation*, Academic Press, New York.
Anderson, J.M. (1992) Response of soils to climate change. *Advances in Ecological Research* **22**, 163–210.
Arden-Clarke, C. and Hodges, R.D. (1988) The environmental effects of conventional and organic/biological farming systems. II Soil ecology, soil fertility and nutrient cycling. *Biological Agriculture and Husbandry* **5**, 223–287.
Boeringa, R. (1980) *Alternative Methods in Agriculture, Developments in Agricultural and Managed-Forest Ecology 10*, Elsevier Publishing Company, Amsterdam.
Bouwman, A.F. (1990) *Soils and the Greenhouse Effect*, John Wiley and Sons, Chichester.
Brown, L.E., Kane, H. and Ayres, E. (1993) *Vital Signs. The Trends that are Shaping our Future*, Earthscan Publications, London.
Dorin, J.W., Molina, J.A.E. and Harris, R.F. (1994) *Defining Soil Quality for a Sustainable Environment*, Soil Science Society of America, Madison.
Ehrlich, A.H. and Ehrlich, P.R. (1987) *Earth*, Methuen London Ltd, London.
Giller, K.E. and Wilson, K.J. (1991) *Nitrogen Fixation in Tropical Cropping Systems*, C.A.B. International, Wallingford.
Greenland, D.J. and Szabolcs, I. (1994) *Soil Resilience and Sustainable Land Use*, C.A.B. International, Wallingford.
Hance, R.J. (1987) Herbicide behaviour in the soil, with particular reference to the potential for ground water contamination. In *Herbicides*, eds. Hutson, D.H. and Roberts, T.R., John Wiley and Sons, Chichester, pp. 223–247.

Harper, J.L. and Hawksworth, D.L. (1994) Biodiversity: measurement and estimation. *Phil. Trans. R. Soc. Lond. B* **345**, 5–12.

Harrison, R.M. (ed) (1992) *Understanding our Environment: An Introduction to Environmental Chemistry and Pollution*, The Royal Society of Chemistry, Cambridge.

Hayes, W.J. and Lawes, E.R. (1991) *Handbook of Pesticide Toxicology Volume 1 General Principles*, Academic Press, New York.

Hayward, J.A. and McChesney, I. (1992) In *Proceedings of the International Conference on Sustainable Land Management*, Napier, Hawke's Bay, New Zealand, 17–23 November 1991, ed. P.R. Henriques, Hawke's Bay Regional Council, Napier, New Zealand, pp. 36–41.

Huston, M. (1993) Biological diversity, soils and economics. *Science* **262**, 1676–1680.

Jacks, G.V. and Whyte, R.O. (1939) *The Rape of the Earth. A World Survey of Soil Erosion*, Faber and Faber Ltd., London.

Kaiser, J.P. and Bollag, J.M. (1990) Microbial activity in the subsurface. *Experentia* **46**, 797–806.

Kang, B.T. (1988) Nitrogen cycling in multiple cropping systems. In *Advances in Nitrogen Cycling in Agricultural Ecosystems*, ed. Wilson, J.R., Commonwealth Agricultural Bureau, Wallingford, pp. 333–348.

Kelley, K.W. (1988) *The Home Planet*, Addison Wesley, Reading.

Leach, G. (1975) *Energy and Food Production*, International Institute for Environment and Development, London.

McElroy, M.B. (1994) Climate of the earth: an overview. *Environmental Pollution* **83**, 3–21.

Nye, P.H. and Greenland, D.J. (1960) *The Soil under Shifting Cultivation*, Commonwealth Agricultral Bureau, Wallingford.

O'Riordan, T. and Cameron, J. (1994) *Interpreting the Precautionary Principle*, Earthscan Publications, London.

Pedersen, K. (1993) The deep subterranean biosphere. *Earth-Science Reviews* **34**, 243–260.

Pfeiffer, E. (1947) *The Earth's Face. Landscape and its Relation to the Health of the Soil*, Faber and Faber Ltd., London.

Rice, E.L. (1984) *Allelopathy*, Academic Press, New York.

Shaxson, T.F., Hudson, N.W., Sanders, D.W., Roose, E. and Moldenhauer, W.C. (1989) *Land Husbandry. A Framework for Soil and Water Conservation*, Soil and Water Conservation Society, Ankeny, Iowa, US.

Thurman, E.M. (1985) *Organic Geochemistry of Natural Waters*, Martinus Nijhoff/Dr W. Junk Publishers, Dordrecht.

Wilson, E.O. (1992) *The Diversity of Life*, W.W. Norton & Co. Ltd., London.

Wood, M. (1991) Biological aspects of soil protection. *Soil Use and Management* **7**, 130–136.

Young, A. (1989) *Agroforestry for Soil Conservation*, C.A.B. International, Wallingford.

Index